HUMAN ANATOMY

HUMAN ANATOMY

RUTH ASHLEY
University of Michigan

in consultation with
JAMES A. McNAMARA, D.D.S., Ph.D.
Department of Anatomy and
Center for Human Growth and Development
University of Michigan

John Wiley & Sons, Inc.
New York • London • Sydney • Toronto

Editors: Judy Wilson and Irene Brownstone
Production Manager: Ken Burke
Editorial Supervisor: Winn Kalmon
Composition and Make-up: Ikuko Workman
Artist: Jim Schulz

Library of Congress Cataloging in Publication Data

Ashley, Ruth
Human anatomy.

(Self-teaching guides)
Bibliography: p.
Includes index.
1. Anatomy, Human—Programmed instruction. 2. Human Physiology—Programmed instruction. I. McNamara, James A. II. Title. (DNLM1. Anatomy. QF4 A826h)
QP31.2.A83 611 76-65
ISBN 0-471-03508-4

Printed in the United States of America.

76 77 10 9

To the Reader

Human anatomy is one of the most relevant topics for study by anyone today. When you know the components of your body, you are able to understand better how and why it functions. You are able to communicate more effectively with health care people should the need arise. You will be able to make more sense out of newspaper and magazine reports on current health developments. And there are many other practical reasons for acquiring a basic knowledge of human anatomy. You may be studying general biology or zoology; the human is similar to other animals in its structural and function and is often included in such courses. You may be working in a health-related field; a knowledge of human anatomy and its vocabulary can be very helpful in making you a productive member of the health care team.

This book cannot give you the complete story of anatomy. But it can give you an overview of the human structure. You will become familiar with both gross anatomy (seen by the unaided eye) and microscopic anatomy, in addition to being introduced to physiology or function. You will have a working knowledge of human anatomy for use in your daily life, as well as a basis to build on, should you decide to pursue the subject further.

Many persons, some anatomists and some not, have contributed to produce this Self-Teaching Guide. If I tried to list them all, the result would be a compilation of most of my meaningful contacts over the past ten years. Nevertheless, I hereby officially thank them all. I specifically wish to express my appreciation for their input and patience to Dr. Jim McNamara, Jim Schulz, and Irene Brownstone.

February 1976 — Ruth Ashley
Ann Arbor, Michigan

How to Use This Book

This Self-Teaching Guide is divided into fifteen chapters, each dealing with a different aspect of human anatomy. Each chapter begins with a list of objectives, which outline what you will learn in studying the chapter; the objectives tell you what you will be able to do when you have completed it. A self-test at the end of each chapter allows you to test yourself, to see if you have indeed learned the material.

Between the objectives and the self-test, each chapter is divided into many smaller numbered segments called frames. Each frame presents new information, followed by several questions about the new material. The correct answers to these questions follow a dashed line near the end of the frame. As you work through this Self-Teaching Guide, use an index card or folded paper to mask the printed answers until you have written your own. After you have answered all the questions in a frame, compare your answers with those given. You need not use the exact words, of course, but be sure you understand any discrepancies before going ahead to the next frame.

Each chapter will take from one to three hours to complete. Whenever possible, try to complete a chapter in one or two study sessions. Stop only at the end of a chapter or a section within a chapter. You will learn the material more easily if your work is not interrupted too frequently.

Take the test at the end of each chapter after you have completed the chapter. Then compare your answers to those given following the self-test. After each answer, the number of the frame or frames in which the information was presented is given in parentheses. If you should miss as many as half of the questions in any self-test, you may find it useful to repeat the chapter before going on. At the end of the book, you will find a final test. This test will help to determine whether you have retained the material you learned as you worked through the Self-Teaching Guide. As before, frame references are given in case you wish to review.

The Index at the back of the book includes a pronunciation guide to basic human anatomy terms. You should find this special feature useful for study, review, and reference.

Prerequisites

You should be able to master the material in this Self-Teaching Guide without any specific background in the area. A high school education, or its equivalent in life experiences, will have provided you with an adequate vocabulary and sufficient motivation to learn the basics of human anatomy through a study of this Self-Teaching Guide.

Cross-Reference Chart for Selected Anatomy and Physiology Texts

You may find the information in the following cross-reference chart useful if you wish to pursue certain topics in greater depth or if you are a student in a course using one of the books listed below. The chart correlates chapters in this book with material on the same subjects in a number of the most widely used anatomy or anatomy and physiology textbooks. The numbers listed in the chart correspond to the chapter numbers in the textbooks.

Anthony, Catherine Parker and Kolthoff, Norma Jane, Textbook of Anatomy and Physiology, ninth edition (St. Louis: Mosby, 1975).

Chaffee, Ellen E. and Greisheimer, Esther M., Basic Physiology and Anatomy, second edition (Philadelphia: Lippincott, 1969).

Crouch, James E., Functional Human Anatomy, second edition (Philadelphia: Lea & Febiger, 1972).

Crouch, James E. and McClintic, J. Robert, Human Anatomy and Physiology, second edition (New York: John Wiley, 1976).

Francis, Carl C. and Martin, Alexander H., Introduction to Human Anatomy, seventh edition (St. Louis: Mosby, 1975).

Gardner, Weston D. and Osburn, William A., Structure of the Human Body, second edition (Philadelphia: Saunders, 1973).

Jacob, Stanley W. and Francone, Clarice Ashworth, Structure and Function in Man, third edition (Philadelphia: Saunders, 1974).

Langley, L. L., Telford, Ira A., and Christensen, John B., Dynamic Anatomy and Physiology (New York: McGraw-Hill, 1974).

McClintic, J. Robert, Basic Anatomy and Physiology of the Human Body (New York: John Wiley, 1975).

Tortora, Gerard J. and Anagnostakos, Nicholas Peter, Principles of Anatomy and Physiology (San Francisco: Canfield Press, Harper & Row, 1975).

Ashley chapter	Anthony/ Kolthoff	Chaffee/ Greisheimer	Crouch	Crouch/ McClintic	Francis/ Martin	Gardner/ Osburn	Jacob/ Francone	Langley/ Telford/ Christensen	McClintic	Tortora/ Anagnostakos
1	1	1	2	1	1, 17	1	1	1	1	1, 6
2	2, 3	2	3, 5	2, 6	2	1, 2	2, 3	2, 3	2	3, 4
3	5	3	7, 8, 9	7	3	3	5	5	6	7, 8
4	6	4, 5	10, 11	9, 10	5	4	7	6	8	10, 11
5	7, 8	6, 7, 8	18	11, 12, 13	6	5	8	8, 9	10, 11, 12	12, 13
6	10	10	19	16, 17, 18	7	5	9	12, 13	15	14
7	11	19	17	31	8	10	15	31	26	15
8	13	11, 12, 13	12	22, 24, 25	9	6	10	15, 16, 17	17, 19, 20	16, 17
9	13	11, 13	12	22, 25	10	6	11	17	17, 20	16
10	12	15	14	26	11	7	12	20, 21	21	19
11	14	16	13	27	12	8	13	24	22	20
12			6	6	14	11	4		5	5
13	16	18	15	29	13	9	14	28	24	22
14	18	20	16	30	16	9	17	32, 33	25	24
15	19	20	16	30	15	9	17	32, 33	25	24

Contents

HUMAN ANATOMY

PART I
Body Organization

In this book, you will find out what comprises a normal human body. You will learn how the organs are constructed and how they function to keep the body operating as an entity. Before we investigate the separate organs, we must first look at the body as a whole. In Chapter 1, we will identify surface landmarks for reference and introduce various anatomical terms for use in later discussions. Then, in Chapter 2, we will examine briefly the basic unit of life, the cell. Millions of cells in communication make up the body. We shall see how the cells are organized into four basic types of tissues, and how the tissues are organized into organs such as the heart, brain, or liver. The organs then are combined into systems, which shall be examined separately in Chapters 3 through 15. When you have finished studying this Self-Teaching Guide, you will be prepared to study further if you wish, or to live with fuller awareness of your body.

CHAPTER ONE
The Human Body

The human body is familiar to all of us, but, as seen from an objective point of view, it is a complex structure. Internally, it is even more complex with many parts interacting for an effectively functioning organism. On the microscopic level, the complexity is so intense that it almost seems unreal. In this chapter, we will look at the body as a whole. We will examine the external landmarks you can see in a mirror or on another person. You will learn some terminology that will make discussions in later chapters easier to understand. Specifically, when you have completed your study of this chapter, you will be able to:

- describe the "anatomical position";
- identify common names of surface structures from anatomical descriptions;
- locate surface structures using the following terms: anterior, posterior, superior, inferior, medial, lateral, proximal, and distal;
- specify the direction of a median, transverse, and coronal plane of section;
- differentiate between a longitudinal and a cross section of an organ;
- list the three major cavities of the human body;
- identify three sub-cavities and their contents within the thoracic cavity;
- name the four quadrants of the abdomen, and the point that is common to all.

ANATOMICAL POSITION

1. The human body can be placed in many different positions. One of the major tasks in anatomy, however, is to specify the location of organs in relation to one another. In order to make this task easier, anatomists always consider the body and its parts in a constant relationship,

which is the anatomical position as shown at the right. From the drawing, you will be able to identify the features critical to the anatomical position.

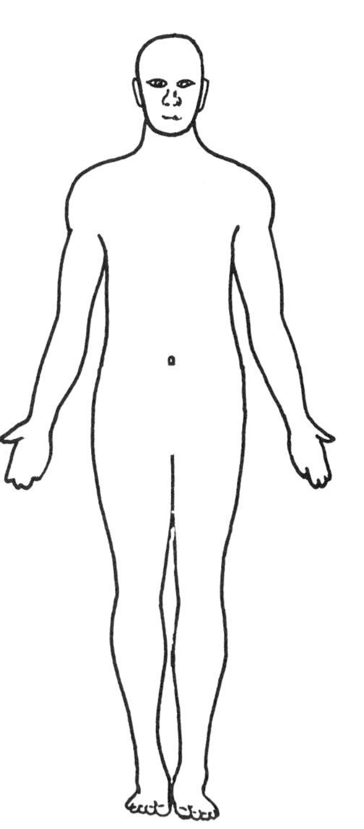

(a) Is the anatomical position standing, sitting, or lying? ______________

(b) What is the position of the eyes?

____

(c) Where are the arms positioned?

____

(d) How are the palms facing?

____

(e) How would you describe the position of the feet? ______________

- - - - - - - - - - - - - - - - - -

(a) standing; (b) even, open, and forward; (c) at sides; (d) forward; (e) parallel and close together

2. When a body is placed in the anatomical position, it is erect. The eyes are open and level; the head is not tipped. The arms are down at the sides, and the palms face forward. The feet are parallel to each other and the heels are close together. When you assume this position, your internal organs and external structures have a specific relationship to each other. You can say, for example, that your chin is positioned lower on your body than your nose, or that your chest is higher than your legs. Consider a body in the anatomical position.

(a) Which are closer together—knees or elbows? ______________

(b) Do the knuckles of the fingers face forward or backward?

____

(c) Is the thumb of the right hand on the right or left of the hand?

____

- - - - - - - - - - - - - - - - - -

(a) knees; (b) backward; (c) right

3. Anatomical relationships are always described as if the body were in the anatomical position. The terminology of the descriptions is more precise than "above" and "below," or "front" and "back." Read the following definitions.

Superior: above, or toward the head.
Inferior: below, or toward the feet.
Anterior: front, or in front of.
Posterior: back, or in back of.

The neck would be described as inferior to the head, and superior to the chest. The nose is on the anterior part of the head, and superior to the mouth. The nostrils, however, are posterior to the tip of the nose.

(a) Locate the elbow in relation to the:

shoulder ____________________

wrist ____________________

(b) Is the elbow on the anterior or posterior side of the body?

____

(c) Locate the chin in relation to the:

mouth ____________________

neck ____________________

chest ____________________

forehead ____________________

(d) Does the drawing in frame 1 show an anterior or a posterior view?

____

- - - - - - - - - - - - - - - - - -

(a) inferior to the shoulder, superior to the wrist; (b) posterior to the side; (c) inferior to the mouth, superior to the neck, superior to the chest, inferior to the forehead; (d) anterior view

4. Structures in the body have other relationships than anterior or posterior and superior or inferior. Four more definitions are given below.

Medial: toward the midline (The midline is an imaginary line running down the center of an anterior or posterior view).
Lateral: away from the midline; toward the side.
Proximal: toward the point of attachment of a part.
Distal: away from the point of attachment of a part.

The eyes are superior to the nose, as well as lateral to it. The fingers are distal to the wrist, which is proximal to them and to the palm. Refer to the drawing in frame 1 if necessary to answer the following.

(a) How would you locate the breast region in relation to the shoulder?

____

(b) Use three terms to locate the ears in relation to the chin.

____

(c) Which toe is in a more lateral position than the others? __________ __________ Which is more medial? ______________________

(d) What structure in the midline is medial and inferior to the breasts, medial and superior to the inguinal or groin region? __________

(e) Which is distal to the other, the ankle or the calf? __________

- - - - - - - - - - - - - - - - - -

(a) inferior and medial; (b) superior, lateral, and posterior; (c) little toe, largest toe; (d) navel; (e) ankle

5. The drawing in this frame shows the body in the anatomical position, from a posterior view.

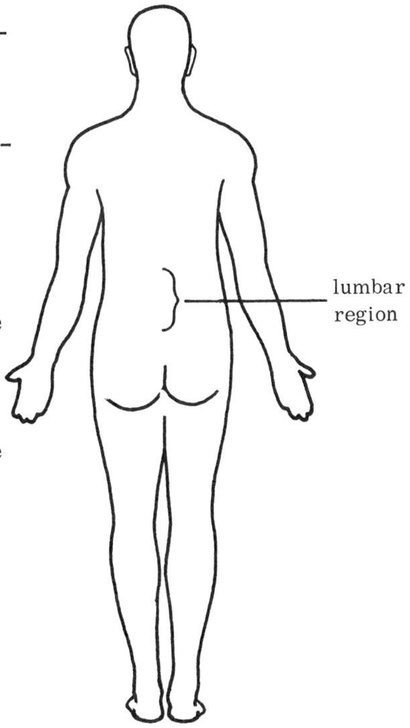

(a) How would you describe the lumbar region (the "small of the back") in relation to the:

hip ______________________

navel ______________________

(b) What structure is located distal to the wrist and lateral to the index finger?

____

(c) What structure is located distal to the hip but proximal to the knee?

____

(d) In what position is the spine relative to the:

head ______________________

knees ______________________

lumbar region

navel ____________________

sides of the body ____________________

- - - - - - - - - - - - - - - - - - -

(a) superior to the hip, posterior to the navel; (b) thumb; (c) thigh; (d) inferior to the head, superior to the knees, posterior to the navel, medial to the sides of the body

PLANES OF SECTION

6. In discussing the human body, it is often helpful to consider imaginary cuts, or planes of section, through the body in different directions. One such plane is the midsagittal or median plane along the midline. This plane neatly divides the body into two halves, right and left. All paired parts, such as eyes and limbs, are separated with the body halves in a median section. A coronal section divides the body into an anterior and a posterior portion. A coronal section does not divide

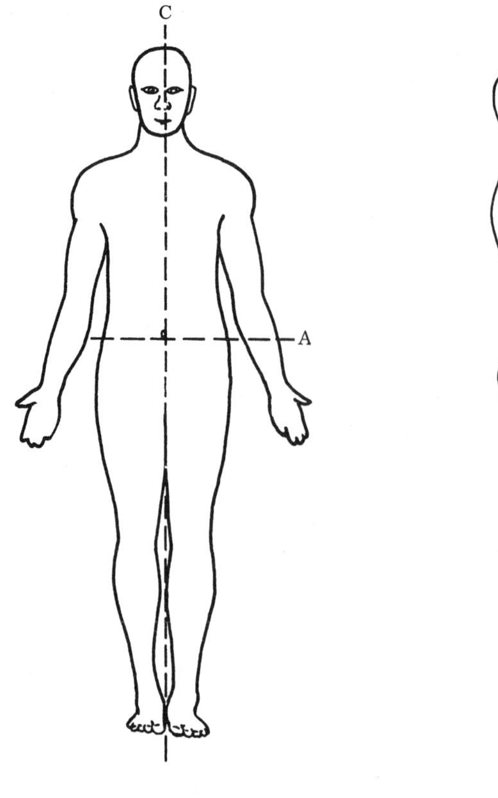

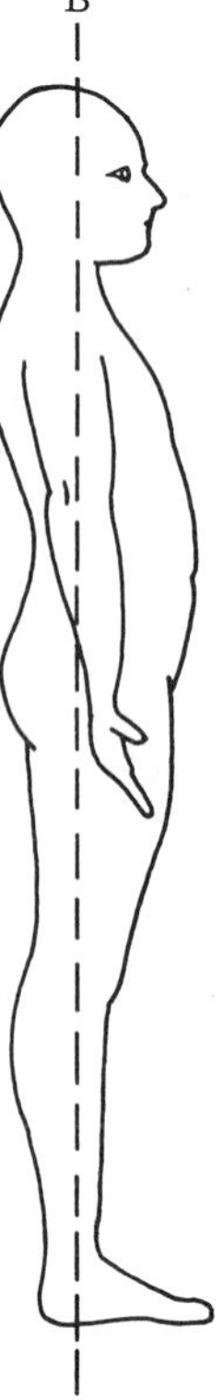

paired parts symmetrically as a median section does. A section crosswise through a body at any level is called a transverse section. Like the coronal section, the transverse plane cannot produce two equal halves.

(a) Which of the lettered planes in the drawing represents each of the following?

coronal section ____________________

median section ____________________

transverse section ____________________

(b) Which plane divides midline structures, such as the nose, in half?

____

(c) Which section would separate the lower portion of the legs from the body? ____________________

(d) Which plane of section could provide two symmetrical halves? ______________

(e) In the drawing on the right, indicate how each of the three planes of section could divide the head, or explain why it couldn't.

- - - - - - - - - - - - - - - - - -

(a) coronal—B, median—C, transverse—A
(b) median
(c) transverse
(d) median
(e) coronal would be up and down at ears

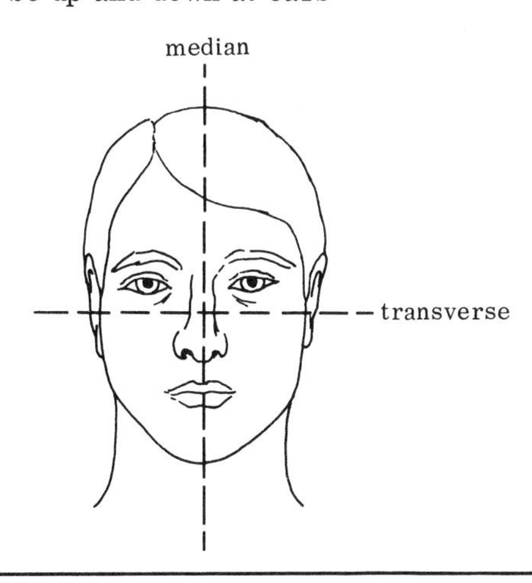

7. The three planes of section are adequate for describing the locations of most structures in the body. When describing the structure of a single organ, however, additional terms are needed. For example, the large bone of the thigh is the femur. It has, as you could guess, a proximal end at the hip and a distal end at the knee. A section through the long axis of the bone is called a longitudinal section. A section through the rounded portion would be called a cross section. Organs and structures that are longer than they are wide are often considered both in longitudinal and in cross section.

   (a) A transverse plane divides the body into superior and inferior portions. A cross section would divide an arm or a leg bone into what two portions? ______________________

   (b) This drawing represents the femur in its anatomical position. A section along the dotted line would be termed ______________________. What two portions would this section divide the femur into? ______________ __________

   (c) In general, which would expose a larger surface—a longitudinal or a cross section of an organ?

   ____

- - - - - - - - - - - - - - - - - -

(a) proximal and distal portions; (b) longitudinal section, medial and lateral portions; (c) longitudinal section

BODY CAVITIES

8. A median plane through the body reveals three closed cavities: the dorsal, the thoracic, and the abdominal. The dorsal cavity includes the cranial cavity and the vertebral canal. The cranial cavity in the head extends in a narrow canal down the back, as the vertebral canal. The word cranial always refers to the head or to the bones surrounding the cranial cavity. The vertebral canal passes through the vertebral column, made up of many separate vertebrae, or back bones. The cranial cavity and vertebral canal house and protect the brain and spinal cord.

   The thoracic cavity is located in the chest. As the drawing on the following page shows, it is anterior to the vertebral canal. When viewed from a coronal section, the thoracic cavity is seen to be divided into three sub-cavities, one in the midline for the heart, and two lateral

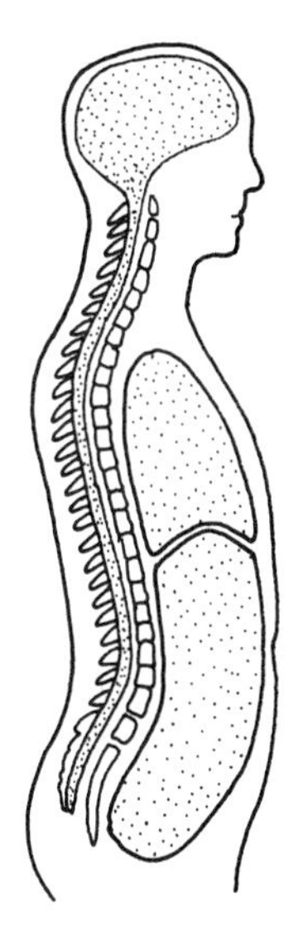

sub-cavities for the right and left lungs. The abdominal cavity also extends laterally, but its sub-cavity, called the pelvic cavity, is located at the inferior portion of the abdominal cavity. The dividing line between the abdominal cavity and the pelvic sub-cavity is not as definite as the divisions in the cranial and thoracic cavities. Later, as you study bones and internal organs, you will discover the landmarks of the pelvic cavity.

Examine the drawing below.

(a) Name the cavity shown by dashed lines.

____

(b) Describe the two divisions of the dorsal cavity in relation to the other two major cavities.

____

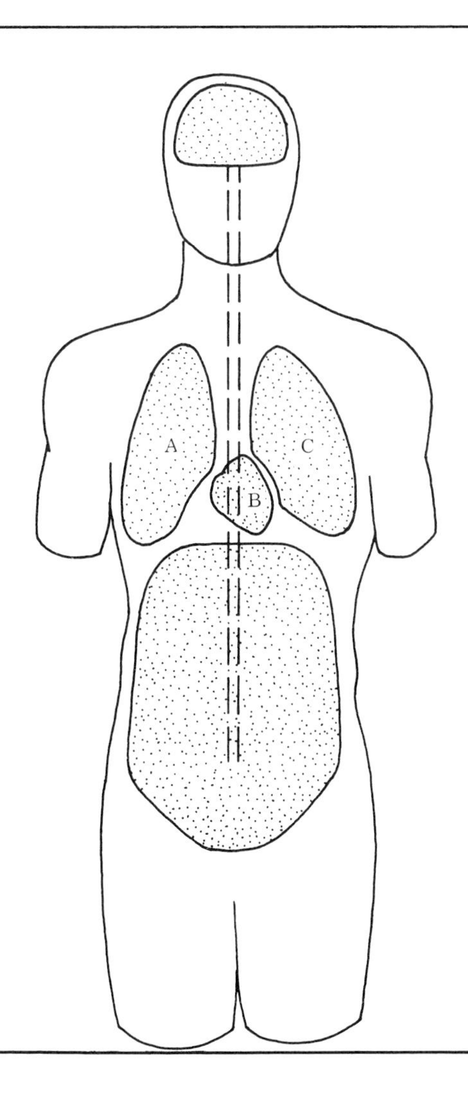

(c) Three cavities are shown in the chest region. Taken together, what are these called? ______________ What structures are housed in the lettered areas of the diagram?

A ______________

B ______________

C ______________

(d) Name the largest cavity in the drawing. ______________

What is the inferior portion of this cavity called? ______________

(e) Describe area C in relation to the:

vertebral canal ______________

navel ______________

cranial cavity ______________

thoracic cavity ______________

- - - - - - - - - - - - - - - - - -

(a) vertebral canal
(b) cranial cavity is superior to thoracic and abdominal cavities; the vertebral canal is posterior to both and part of it is superior to them.
(c) thoracic cavity; A—right lung; B—heart; C—left lung
(d) abdominal cavity; pelvic cavity
(e) anterior and lateral
superior and posterior (internal is as good as posterior here)
inferior and lateral
makes up left lateral part of thoracic cavity

9. Many internal organs are located in the abdominal and pelvic cavities. These include most digestive organs that process the food we eat, and urinary organs that manufacture urine and regulate water and chemical balance for the body. Some sex organs and glands that produce substances needed for life are also found in this large cavity.

In order to make locations more specific, we mentally divide the abdomen into four areas or quadrants using the navel as a point of intersection. Then organs are referred to as being in the upper right quadrant, the lower left quadrant, and so on. These reference lines will be useful when we get into the abdominal organs in later chapters. Notice that the right and left sides appear as they do in your own or another person's body. Thus the anatomical right always appears on the left of the drawing. This pattern is consistent throughout this and other anatomy texts.

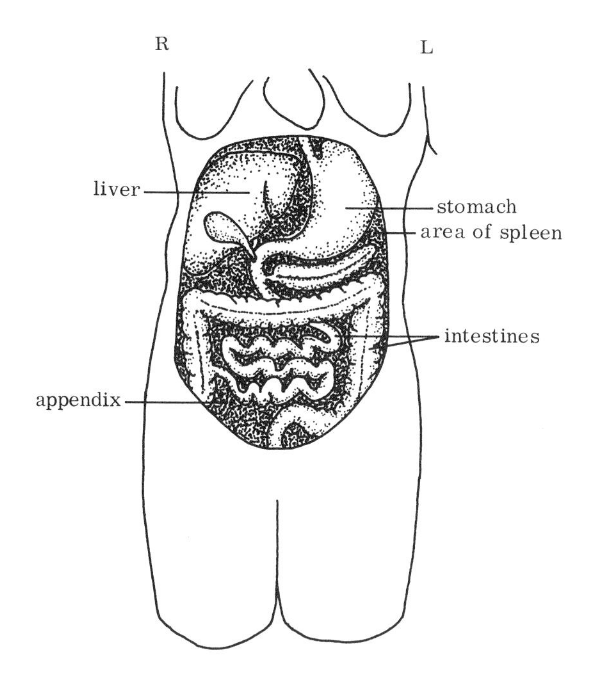

(a) In what area of the abdomen is the appendix located? ______________________

(b) In what area of the abdomen is most of the liver located? ______________________

(c) Describe the location of the spleen in general, and relative to the stomach. ______________________

(d) What surface structure is common to all four quadrants? ______________________

(e) Name an organ that is also in all four quadrants. ______________________

(f) From what plane is the drawing in this frame made? ______________________

- - - - - - - - - - - - - - - - - - -

(a) lower right quadrant
(b) upper right quadrant
(c) spleen is in upper left quadrant, lateral to stomach
(d) navel (also skin, of course)
(e) small intestine or large intestine
(f) coronal

10. Indicate the major body cavity that houses each organ below.

 (a) brain ______________________

 (b) spinal cord ______________________

 (c) lungs ______________________

 (d) heart ______________________

 (e) digestive organs ______________________

- - - - - - - - - - - - - - - - - -

(a) cranial cavity; (b) vertebral canal; (c) thoracic cavity (lateral sub-cavities); (d) thoracic cavity (midline sub-cavity); (e) abdominal cavity

SELF-TEST

This Self-Test is designed to show how well you have mastered this chapter's objectives. Answer each question to the best of your ability. Correct answers and review instructions are given at the end of the test.

1. How are the torso, arms, and legs arranged in the anatomical position?

____

____

2. Where is the thumb located in relation to the wrist? ______________

____

3. Where is the neck located in relation to the shoulder? ______________

____

4. Name three surface structures through which a median plane of the body passes. ______________________________

____

5. Name the planes indicated by the dashed lines.

   A ____________________

   B ____________________

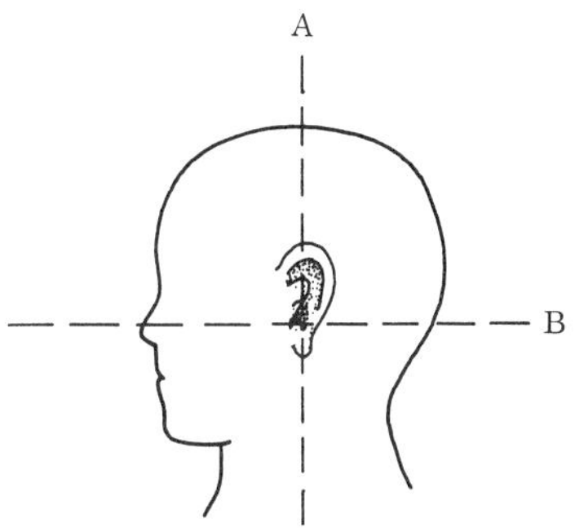

A

B

6. The drawing at the right represents an abdominal organ called the pancreas. Mark on it the direction of: a longitudinal section (A) and a cross section (B).

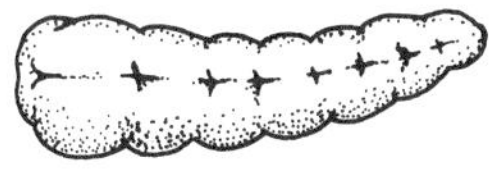

7. What body cavity encloses the brain and the spinal cord? __________

____

8. Name three organs housed in three sub-cavities in the chest. _______

____

9. Describe the location of the pelvic cavity. Which quadrant is it in?

____

____

10. In which quadrant is the navel located? ____________________

Where is it located in relation to a breast? ____________________

____

Answers

Compare your answers to the Self-Test questions with those answers given below. If all of your answers are correct, you are ready to go on to the next chapter. If you missed any, review the frames indicated in parentheses following the answers. If you missed several questions, you should probably reread the entire chapter carefully.

1. trunk erect; arms at sides with palms forward; legs together with feet parallel (frame 1)
2. distal and lateral (frame 4)
3. superior and medial (frame 5)
4. nose, mouth, navel (or any surface structures on midline) (frame 6)
5. A—coronal; B—transverse (frame 6)
6. 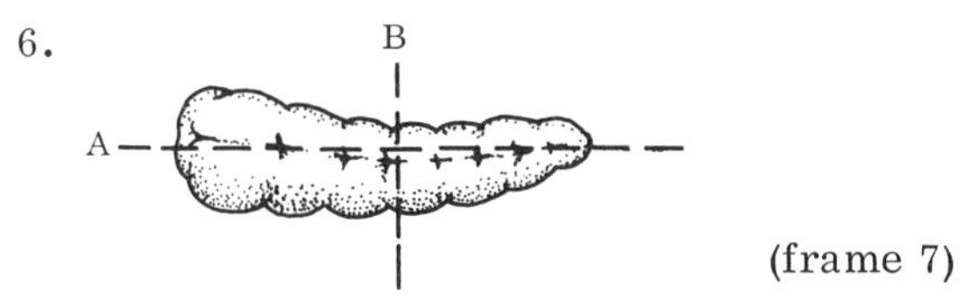

(frame 7)
7. dorsal cavity (cranial cavity and vertebral canal) (frame 8)
8. right lung, left lung, heart (frame 8)
9. inferior part of abdominal cavity; lower right and left quadrants (frame 9)
10. all, since it is the point of intersection; inferior and medial to breast (frame 9)

CHAPTER TWO
Cells and Tissues

The human body is made up of microscopic living cells, each independently carrying out its life functions. In this chapter we will examine the characteristics of a typical cell, then look more closely at the cells that make up two of the four body tissues. When you complete your study of this chapter, you will be able to:

- label the structures in a typical cell;
- specify the function of various organelles;
- list in order the stages of mitosis, giving one distinguishing feature of each phase;
- differentiate between epithelial and connective tissues;
- classify epithelial tissue according to cell shape and arrangement;
- explain how epithelial tissues are joined together;
- identify variations of epithelial cells from descriptions;
- state the functions of three connective tissue cells;
- describe how cartilage differs from general connective tissue.

CELLS

The human body is made up of millions of separate, individual cells. These cells are not all alike in structure, nor are they necessarily similar in function. But cells in general carry out the same processes as do all larger living things. Cells respond to their surroundings, they undergo respiration, they digest things, they dispose of waste materials. In this section we will examine the basic structure of a cell and how it carries out its functions.

1. A typical cell has two main divisions, one within the other, each enclosed in a membrane. The outer membrane is called the cell membrane

or plasma membrane. Enclosed within it are both the cytoplasm and the nucleus of the cell. A nuclear membrane also surrounds the nucleus, effectively segregating it from the cytoplasm.

The nucleus is the control center for the cell. It directs all activities that are carried on in the cytoplasm. In the nucleus is genetic material (chromatin) that is reproduced whenever the cell divides to produce two identical smaller cells. Specific large molecules are put together in the nucleus, which then move through pores or openings in the nuclear membrane to carry instructions to the cytoplasm.

(a) Identify the structures indicated in the diagram at the right.

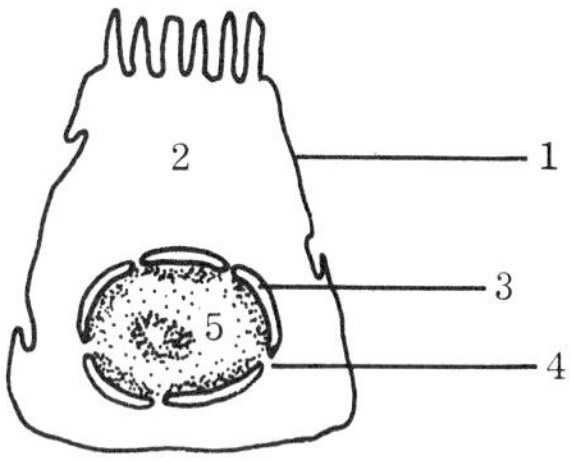

1 ____________________

2 ____________________

3 ____________________

4 ____________________

5 ____________________

(b) What passes through structure 4? ____________________

(c) In what part of the cell is genetic material found? ____________

What is this material called? ____________________

(d) Which major portion of the cell controls the activities of the other part? ____________________

- - - - - - - - - - - - - - - - - -

(a) 1—plasma (cell) membrane
2—cytoplasm
3—nuclear membrane
4—pore or opening
5—nucleus

(b) large molecules (special ones)

(c) nucleus; chromatin

(d) nucleus

2. Several different types of organelles (little organs) are located in the cytoplasm of a cell. Each of these has a different structure and function. The typical cell is something like a factory, with the nucleus performing the function of the executive office. In the cytoplasm, different organelles perform the functions of manufacturing various products, packaging them, transporting them, providing energy for all this, and disposing of any waste or by-products.

The manufacturing of cellular products takes place in the endoplasmic reticulum, a network of membranes that enclose a space where products are held. This endoplasmic reticulum (abbreviated ER) may be rough or smooth surfaced; the rough surfaced endoplasmic reticulum (RER) has many ribosomes attached to the membrane that are active in protein production. Many of the large molecules that passed through nuclear pores, along with proteins produced in the cytoplasm, make up these ribosomes. As more products are produced, the RER fills up, and bits are pinched off at ends. The bits move toward the Golgi apparatus, which removes some fluid as the substance is packaged. The Golgi apparatus resembles a stack of pancakes. Carbohydrates may be added to the product as it is packaged into rounded structures called condensing vacuoles. Here the product becomes even more concentrated as it is transported toward the plasma membrane. The cell may produce several different products, but all go through this general process.

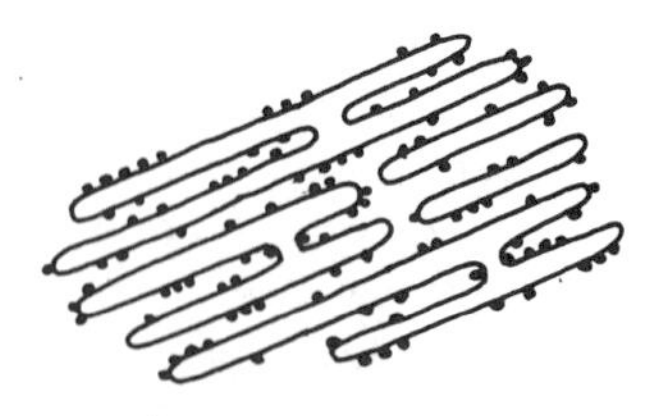

Lysosomes are also present in a typical cell; these organelles digest any waste material that may result, as well as acting to rid the cell of toxins or viruses that may enter. The energy for all this is provided by mitochondria—small double-membraned structures in which chemical reactions resulting in an excess of energy take place.

(a) The cell can be compared to a factory. Name the organelle that serves each of the functions below.

(1) energy supply ______________________

(2) executive office ______________________

(3) disposal of waste ______________________

(4) production machinery ______________________

(5) transport of products ______________________

(6) packaging of products ______________________

(b) What structures are present on RER? ______________________

- - - - - - - - - - - - - - - - - -

(a) (1) mitochondria; (2) nucleus; (3) lysosomes; (4) endoplasmic reticulum (ER); (5) condensing vacuoles; (6) Golgi apparatus

(b) ribosomes

3. Name the organelles diagrammed in the following figures.

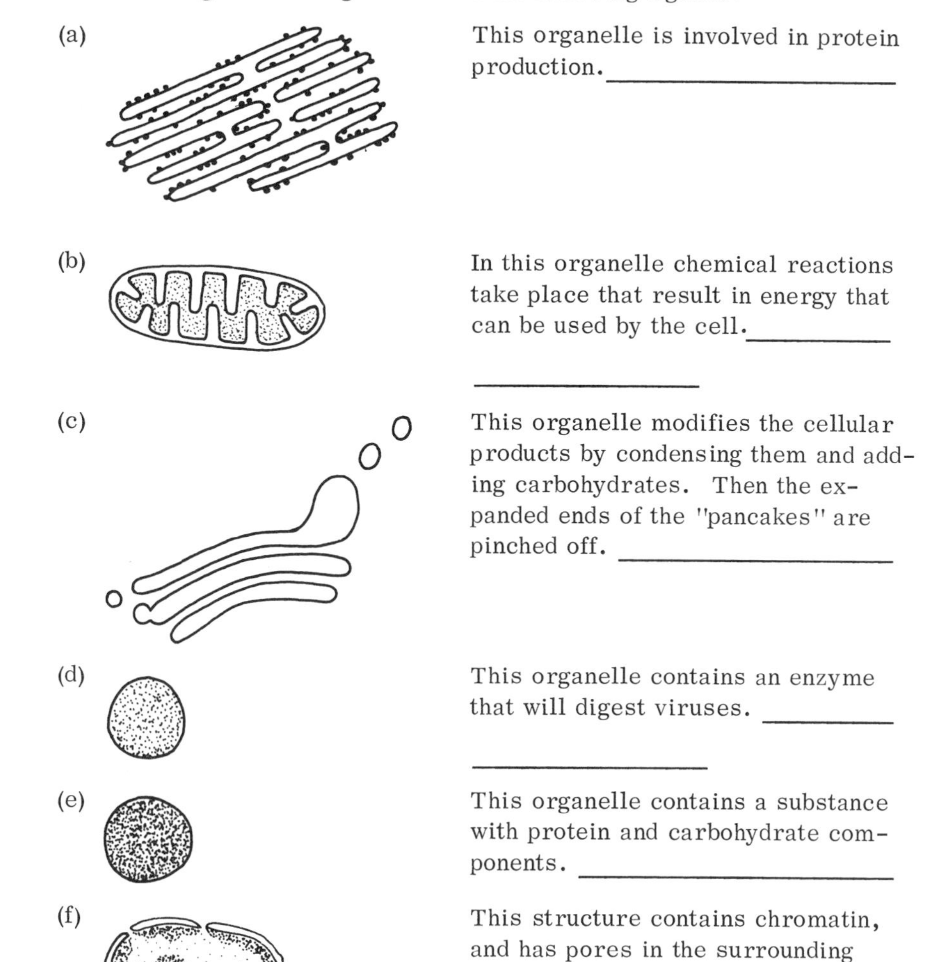

(a) This organelle is involved in protein production.____________________

(b) In this organelle chemical reactions take place that result in energy that can be used by the cell.__________ ________________

(c) This organelle modifies the cellular products by condensing them and adding carbohydrates. Then the expanded ends of the "pancakes" are pinched off. ____________________

(d) This organelle contains an enzyme that will digest viruses. __________ ________________

(e) This organelle contains a substance with protein and carbohydrate components. ____________________

(f) This structure contains chromatin, and has pores in the surrounding membrane. ____________________

\- - - - - - - - - - - - - - - - - -

(a) rough-surfaced endoplasmic reticulum (RER); (b) mitochondria; (c) Golgi apparatus; (d) lysosome; (e) condensing vacuole; (f) nucleus

4. Most of the organelles described in frame 3 are found in most of the cells of the body. The size, the quantity, and the complexity of organelles may vary from cell to cell. A mature cell form may not contain any of a specific organelle, or it may contain so many of that type that it seems full.

4. In addition to organelles, other structures may also be found in cells. Microtubules are very tiny tubes that may function in intracellular (within the cell) transport or in forming a cytoskeleton (cyto means cell) to maintain the shape of a cell. Microfilaments are threadlike structures that also aid in forming a cytoskeleton. In addition, substances such as nutrients, fat, or minerals, may be stored in cells.

   (a) Name two structures that may contribute to forming a cytoskeleton.

   ____

   (b) Nerve cells have very long processes, or extensions, from the main body of the cell. A structure within the cell extends from the cell body almost to the end of the process. Manufactured substances pass from the body of the nerve cell down the hollow center of the structure to the end of the process. What is the structure? ______

   ____

   (c) When two cells are attached together, many threadlike structures come together near the point of attachment in each cell. What are these structures? ______________

- - - - - - - - - - - - - - - - - -

(a) microfilaments, microtubules; (b) microtubule; (c) microfilaments

Cell division

5. Most of the cells in the human body divide as growth proceeds. Some cells, such as nerve cells and muscle cells, do not divide, although they may grow in size or length after birth. When cells do divide, the division process is called mitosis. For most of its life, a cell is between divisions; this is called interphase since it represents the time between the phases of mitosis. During interphase, the nucleus contains scattered chromatin. The action of this chromatin is crucial to understanding mitosis. While other changes take place also, we will focus on only a few major changes in the nucleus.

   Mitosis includes several phases: prophase (pro- means beginning), metaphase (meta- means middle), anaphase (ana- means progressing), and telophase (telo- means completion).

   In prophase, 23 pairs of distinct chromosomes form from the scattered chromatin, and the nuclear membrane disappears. During metaphase, the 23 pairs of chromosomes line up in the center of the cell. During anaphase, each pair separates, with one chromosome of each pair being drawn toward opposite ends of the cell. During telophase, the cytoplasm is divided with organelles being approximately equal, new nuclear membranes are formed, and division is complete. Now a new interphase begins. During this period the chromatin will reproduce itself in preparation for the next mitosis.

(a) During which phase of cell division does duplication of chromosomes take place? ____________________

(b) Label the stages of mitosis shown at the right (for simplicity, only 3 pairs of chromosomes are shown).

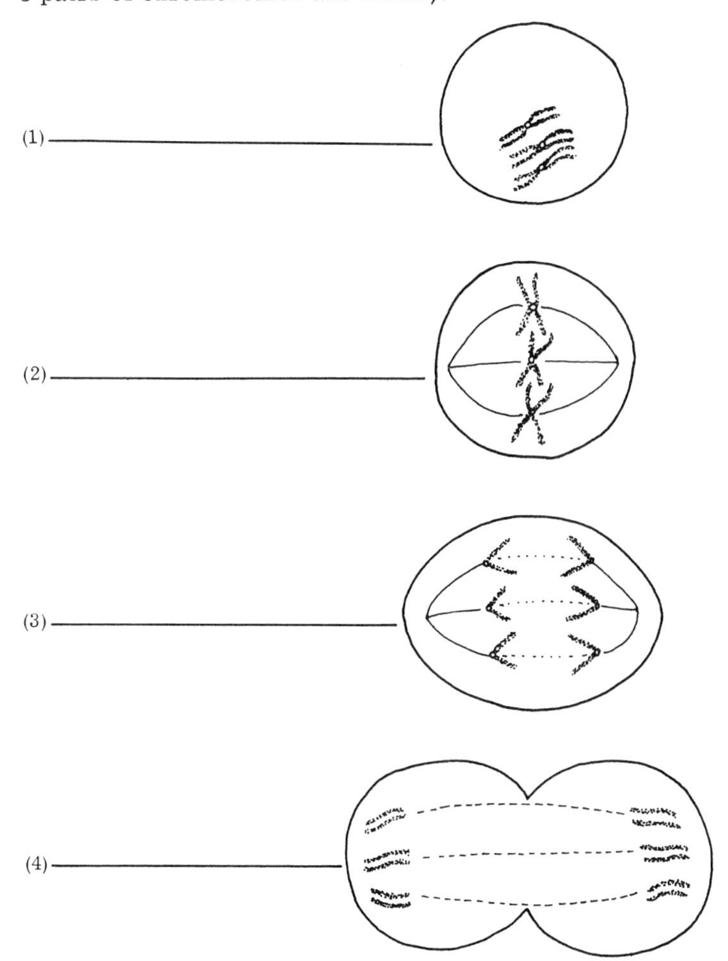

(c) If the original cell contained 50 mitochondria, can you say how many would be in each resulting cell? ____________________

- - - - - - - - - - - - - - - - - - -

(a) interphase
(b) (1) prophase; (2) metaphase; (3) anaphase; (4) telophase
(c) No. Each would contain about half the original number.

6. A cell called a macrophage functions in capturing and digesting foreign materials in its environment.

   (a) What organelle would you expect to be exceptionally numerous in a macrophage? ____________________

   (b) After mitosis, would you expect the two resulting macrophages to each have as many lysosomes as the original one? Why? ________

   ____

- - - - - - - - - - - - - - - - - - -

   (a) lysosomes (see frame 2)
   (b) No. Each would have about half as many since they are divided, not reproduced; more are produced during interphase.

TISSUES

The cells of the body are organized into tissues. Each of the four general groups of tissues is made up of similar cells. These groups are epithelial tissues, connective tissues, muscle tissues, and nervous tissues. In this section, we will discuss epithelial and connective tissues. In later chapters, we will examine muscular (Chapter 4) and nervous tissues (Chapter 6).

Epithelial tissue

7. Epithelial tissues are made up of epithelial cells. These tissues cover the body and line all of its cavities and organs; thus the cells must be closely attached. The close attachment is critical, or bacteria could easily enter the body, or fluids could be easily lost.

   Generally, one side of a sheet of epithelial cells is free, exposed to air or body fluids. The other side of the epithelium rests on connective tissue that is well supplied with blood vessels. None of these blood vessels penetrate the epithelial tissue. The outer layer of the skin, for example, is an epithelial tissue. The outer surface is free, while the base of this epithelium rests on connective tissue containing many blood vessels. (Any tissue containing many blood vessels is said to be vascular.) The cells in the epithelial tissue are very close together, with a very small amount of extracellular (outside the cells) fluid.

   At corresponding points in two adjacent cells, many microfilaments merge into thickenings of the plasma membrane. These areas are called desmosomes, and seem to "snap" the two cells together. At the base of the epithelium, structures resembling half a desmosome (hemidesmosome) appear to "snap" the epithelium to the underlying connective tissue. A layer of epithelial tissue lines the trachea (windpipe) that carries air from the environment to the lungs.

(a) Which surface of the epithelium lining the trachea is adjacent to air—free or attached? ____________________

(b) Where in this epithelial lining would you expect to find desmosomes?

____

(c) Where would you find hemidesmosomes? ____________________

(d) When you scratch your arm lightly, it does not bleed, but when you cut it deeply, blood appears. How would you explain this? ________

____

- - - - - - - - - - - - - - - - - -

(a) free
(b) joining adjacent cells in the epithelial tissues
(c) in cells adjacent to the connective tissue (on the attached surface)
(d) No blood vessels are in the epithelium. Only when a break is deep enough to reach the underlying connective tissue can blood be released.

8. Cells in epithelial tissue appear in several characteristic shapes and arrangements. Three general epithelial cell shapes are flat (called squamous), cuboidal, and columnar. The cells are arranged either in a single layer (called simple) or many layers (called stratified). When cells of more than one shape appear in stratified epithelium, the tissue is named for the shape of the cells on its exposed surface. For example, the epithelium of the skin contains many layers of cells. Some are cuboidal, but those nearer the free surface are flat. Thus the epithelium of the skin is classified as stratified squamous epithelium.

(a) Label the epithelial cells below according to the general shape of the cell.

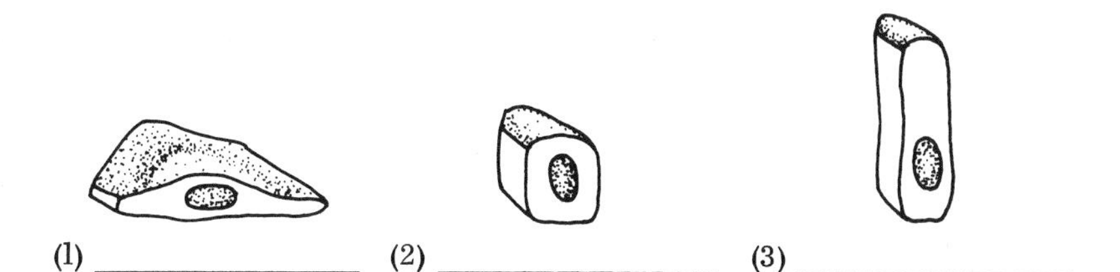

(1) ______________ (2) ______________ (3) ______________

(b) Blood vessels are lined with a single layer of flat cells joined by desmosomes. This layer is surrounded by connective tissue. What type of epithelium lines the blood vessels? ____________________

(c) Part of the epithelium lining the throat is made up of several layers of cells. The cells adjacent to the connective tissue are cuboidal, while those at the free surface are columnar. Intermediate layers are intermediate in height between cuboidal and columnar cells.

How would you classify this epithelium? ______________________

- - - - - - - - - - - - - - - - - -

(a) (1) squamous; (2) cuboidal; (3) columnar (These are all side views of cells. A view from the top of any epithelium gives a rather hexagonal cell whose height can't really be determined.)
(b) simple squamous
(c) stratified columnar

9. Two additional arrangements of epithelium are also seen in the human body: pseudostratified and transitional.

Pseudostratified epithelium often appears to be stratified columnar in type. But actually it is a single layer of epithelial cells of different heights. All of the cells, however, reach the connective tissue underlying the epithelium. The name pseudostratified refers to the false (pseudo means false) layering that appears to be present.

Transitional epithelium is a variation of stratified cuboidal. The shape of these cells varies with stretching of the tissue. The layering varies from about six layers when transitional epithelium is relaxed to only about two or three, somewhat flattened, layers when it is stretched. Very few desmosomes are present in transitional epithelium, which allows the cells to slip over each other and be easily rearranged.

(a) Identify these types of epithelium.

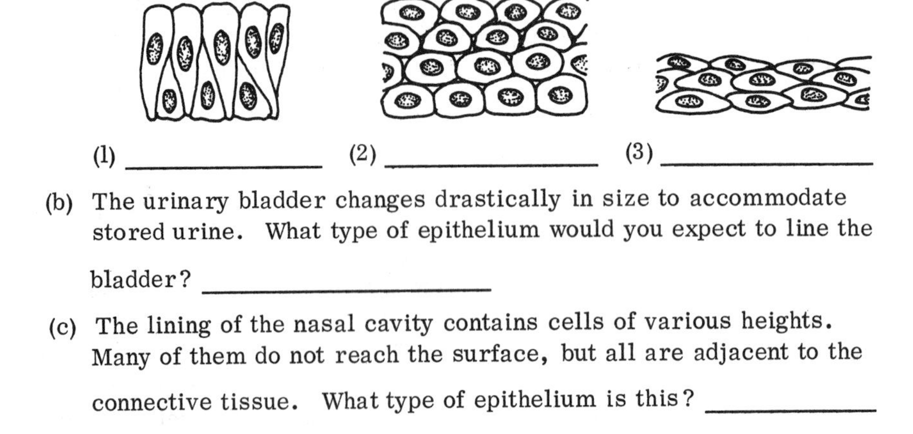

(1) ______________ (2) ______________ (3) ______________

(b) The urinary bladder changes drastically in size to accommodate stored urine. What type of epithelium would you expect to line the bladder? ______________________

(c) The lining of the nasal cavity contains cells of various heights. Many of them do not reach the surface, but all are adjacent to the connective tissue. What type of epithelium is this? ____________ ______________________

(d) The cells lining the stomach are considerably taller than they are wide. Each epithelial cell extends from the connective tissue to the free inner surface of the stomach. What type of epithelium is this? ____________________

- - - - - - - - - - - - - - - - - -

(a) (1) pseudostratified; (2) transitional (relaxed it resembles stratified cuboidal); (3) transitional (stretched it resembles stratified squamous)
(b) transitional
(c) pseudostratified
(d) simple columnar

10. The cells in epithelial tissues frequently show special modifications (specializations) that reflect the function of the specific type or location of the cell. The free surface of the cells may be scaly and dry as in the skin, as a result of a cellular product called keratin. Elsewhere, the surface may have many fingerlike projections to greatly increase the surface area for purposes of absorption. These projections are called microvilli and are very numerous in the epithelial lining of the intestines. Cilia, or hairlike extensions, are found on other epithelial cells, lining most of the respiratory system. These cilia move in a wavelike manner to remove particles from inhaled air. Some epithelial cells are modified into small one-celled glands that secrete mucous to moisten the surface for lubrication or trapping of passing substances. These mucous glands are often called goblet cells because of their shape.

(a) Name the specializations below.

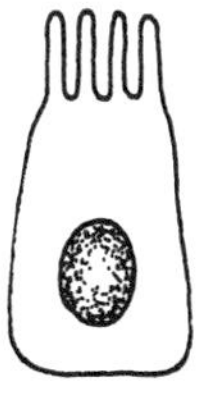

(1) ______________ (2) ______________ (3) ______________

(b) In the kidney, epithelium lined tubules absorb many substances from urine. What specialization of epithelial cells would you expect to find in the kidney? ____________________

(c) In tubes leading toward the lung are two epithelial specializations. One functions to moisten air and trap particles by its stickiness. The other removes particles by sweeping them up away from the lungs in waves. What are these specializations called?

____

(d) The cells on the surface of the skin have a specialization also; they are keratinized. Are these surface cells wet or dry?

_______________ What shape are they? _______________

- - - - - - - - - - - - - - - - - -

(a) (1) cilia; (2) microvilli; (3) goblet cell or mucous gland
(b) microvilli
(c) goblet cells and cilia
(d) dry; flat or squamous

Connective tissue

11. Connective tissue serves to connect, support, and protect parts of the body. It underlies all epithelium, connecting it to other structures. Connective tissue differs from epithelial tissue in several features. Where epithelial tissues have very little extracellular substance, connective tissues have a great deal. The cells of connective tissue are scattered in the extracellular substance, which may also include fibers to give it strength and body. Connective tissues include not only the "basic" connective tissue of the type that underlies epithelium, but also such substances as blood, bone, and cartilage. All of these have the feature of relatively few cells in a large amount of extracellular substance. The structure of bone and blood will be covered in later chapters dealing with the skeletal and circulatory systems.

(a) Which would have a higher proportion of cells, epithelium or its underlying connective tissue? _______________

(b) Which would have a higher proportion of extracellular substance, epithelium or its underlying connective tissue? _______________

(c) Which would contain blood vessels? _______________

- - - - - - - - - - - - - - - - - -

(a) epithelium; (b) connective tissue; (c) connective tissue

12. The fibers found in the extracellular substance of connective tissue are produced by the most common connective tissue cell type, the fibroblast (blast means forming, so the name means fiber-forming). Since they produce fibers, the fibroblasts contain all of the organelles we discussed as being in a typical cell, with much rough-surfaced endoplasmic reticulum (RER).

Another common cell type in connective tissue is the macrophage (phago means to ingest or eat), which ingests particles, such as bacteria or dead cells. These cells are distinguished by large numbers of lysosomes and a "ruffled" or folded plasma membrane that aids in capturing particles.

Several types of blood cells also are commonly found in connective tissue, as are a few important, but uncommon, specialized cells active in immunity and circulation control.

One other cell type is often quite common—the adipose, or fat, cell. The adipose cell is usually almost completely filled with oil, which compresses the nucleus to a flat disk. Some areas of the body are more likely to have masses of adipose cells than others, but all bodies contain some.

(a) What cell produces connective tissue fibers? ____________________

(b) Which cell ingests dead cells or bacteria found in connective tissue?

____

(c) Which cell would be especially numerous in a very large person?

____

(d) What two features could be used to identify a macrophage? ________

____

(e) What shape is the nucleus of an adipose cell? ____________________

- - - - - - - - - - - - - - - - - -

(a) fibroblast; (b) macrophage; (c) adipose; (d) many lysosomes and ruffled cell membrane; (e) flat, compressed by fat or oil

13. Cartilage differs from general connective tissue in that the extracellular substance is fairly firm and provides some rigidity. Cartilaginous fibers are produced by the fibroblast cells. When the cells become trapped in hardening substance, they are called chondrocytes (chondro means cartilage; cyte means cell).

    Unlike general connective tissue, cartilage is not vascular. Nutritional requirements of chondrocytes seem to seep in through the gelatinous substance surrounding them. The cartilage is covered by a perichondrium (peri means around) that forms a fibrous capsule equipped with blood vessels.

    Cartilage is found on the ends of long bones, on the anterior ends of ribs, and in several head and neck structures like the nose, ear, and voice-box or larynx. Most joints also contain some cartilage.

    (a) What is the primary difference between cartilage and general connective tissues? ____

    ____

(b) Where would you find a chondrocyte? ______________________

(c) Is the vascularity of cartilage more like that of general connective tissue or that of epithelium? ______________________

(d) What is the term that describes the fibrous covering of cartilage?

____

- - - - - - - - - - - - - - - - - -

(a) in cartilage the extracellular substance is more rigid
(b) trapped in hardened cartilage
(c) epithelium
(d) perichondrium

Two more tissues are found in the human body. Muscular tissue will be covered in Chapter 4, while nervous tissue is covered in Chapter 6. The four tissues in varying proportions make up all the separate organs of the body. The organs are then organized into systems, each of which has different functions. All the systems are then integrated into the organism—the complete human being. In the rest of this guide, the systems will be treated separately. But keep in mind that they are all functioning together. All systems are needed for maintenance of each body.

SELF-TEST

This Self-Test is designed to show how well you have mastered this chapter's objectives. Answer each question to the best of your ability. Correct answers and review instructions are given at the end of the test.

1. Write the names of the indicated structures.

   A ______________________

   B ______________________

   C ______________________

   D ______________________

   E ______________________

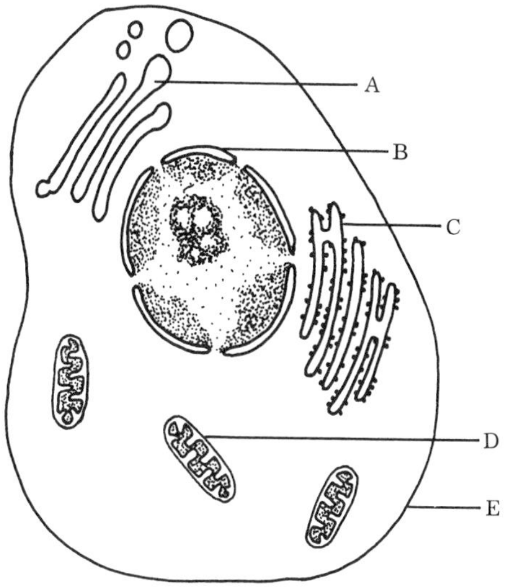

A
B
C
D
E

2. Specify the organelle that performs each of the functions listed below.

   (a) energy production ________

   ____

   (b) intercellular digestion

   ____

   (c) supports the cell or intercellular transport ______________

3. In which stage of mitosis does the division of cytoplasm take place?

   ____

4. Name the structure that joins adjacent epithelial cells. __________

   __________ What other cellular structures are involved in joining cells? ______________

5. The lining of a particular organ consists of epithelial cells. One end of each cell rests on the underlying connective tissue, but not all of the cells reach the free surface. Hairlike structures cover the free surface. Classify the tissue and name the specialization. __________

   ____

6. In a stratified squamous epithelial tissue, why is it possible to have cuboidal cells next to the connective tissue? ______________

   ____

7. Compare epithelial and connective tissues in relation to the proportion of cells to extracellular substance. ________________________________

____

8. Give one function of each of the following cells.

   (a) macrophage ______________________

   (b) fibroblast ______________________

   (c) adipose ______________________

   (d) goblet ______________________

9. What is the primary difference between cartilage and general connective tissue? ____

10. Which of these is more vascular: epithelium, general connective tissue, or cartilage? ____

Answers

Compare your answers to the Self-Test questions with those answers given below. If all of your answers are correct, you are ready to go on to the next chapter. If you missed any, review the frames indicated in parentheses following the answers. If you missed several questions, you should probably reread the entire chapter carefully.

1. A—Golgi apparatus; B—nuclear membrane; C—endoplasmic reticulum; D—mitochondria; E—plasma membrane (frame 1)
2. (a) mitochondria; (b) lysosomes; (c) microtubules (frames 3 and 4)
3. telophase (frame 5)
4. desmosomes; microfilaments (frame 7)
5. pseudostratified with cilia (frame 9)
6. Stratified epithelium is named for the shape of cells in the free surface layer. (frame 8)
7. Connective tissue has relatively fewer cells with much more extracellular substance than epithelial tissue does. (frame 1)
8. (a) ingest particles (frame 12)
   (b) produce fibers (frame 12)
   (c) store fat (frame 12)
   (d) produce mucous (frame 10)
9. In cartilage, the intercellular substance is more rigid and there are no blood vessels. (frame 13)
10. General connective tissue (frames 11 and 13)

PART II

The Framework of the Body

Two major systems of the human body form its framework. The skeletal system provides for support while the muscular system enables the body to move. The bones give the body rigidity or stiffness. Muscles are usually attached to at least two bones, and produce movement by contracting, or shortening. The exact movement produced depends on the location of the muscle attachment. Bone tissue is a specialized form of connective tissue; it too has very few cells in comparison to the rest of its substance. Muscle is made up of a new tissue type. The cells of muscular tissue are somewhat different from those in connective tissue or epithelium. The differences, while small, are critical. Without the small differences on the cellular level, major differences such as those between muscle and bone would not exist. The human body, in fact organized life as we know it, would not exist.

A familiarity with structures in the framework of the body will make it easier for you to grasp the material in the remainder of this guide. Many other body structures are more easily described in terms of their relationships to certain bones and muscles. In addition, a knowledge of the names of individual bones will provide you with a core of anatomical terminology which will be useful to you in future biology studies as well as in coping with your own health and that of your family.

CHAPTER THREE
The Skeletal System

The adult human body contains an internal skeleton which normally includes 206 separate bones. In this chapter, we shall see what makes up a bone and how bones are classified. The names of the bones are important in locating organs to be discussed in later chapters. In addition, you will find the skeletal knowledge valuable in understanding and coping with the many changes that your own body will undergo throughout life. When you complete your study of the chapter, you will be able to:

- list four functions of the skeletal system;
- identify the cells involved in formation of bone;
- classify bones into the four categories based on shape;
- label correctly the bones of the appendicular and axial skeletons;
- differentiate between cranial and facial bones;
- identify the skeletal division to which a given bone belongs;
- name the bones of the cranium, face, and thorax.

BONES IN GENERAL

1. The skeletal system provides a framework of support for the rest of the tissues of the body. Without it, there would be no coordinated movement. The skeletal system also provides protection for many of the organs encased within it. In addition, the skeleton functions as a storage center for various minerals, primarily calcium which makes up a large part of every bone. Some bones, those with marrow cavities, are critical in the production of blood cells which are later released into circulating blood.

   These four functions—support, protection, mineral reserve, and blood formation—are crucial to the maintenance and survival of the body. Several situations involving skeletal elements are described below. Write the specific function of each.

(a) In the central cavities of long bones, a substance called marrow is active in producing cells for release into blood. ______________

(b) The brain is encased in flat bones that serve to prevent injuries to this delicate and vital organ. ______________

(c) The biceps muscle is attached to a bone in the forearm and one in the shoulder; it contracts to bend the elbow. ______________

(d) The calcium stored in bones can be released under the proper conditions and added to the body fluids. ______________

(e) The heart is an organ located in the chest. The bones of the thorax, including ribs and "breastbone," surround the heart, shielding it from outside pressure. ______________

- - - - - - - - - - - - - - - - - -

(a) blood formation; (b) protection; (c) support; (d) mineral reserve; (e) protection

2. The skeletal system, or the skeleton as a whole, provides the four functions of support, protection, mineral reserve, and blood formation. But all bones do not provide all of these functions, although all contribute to support and the mineral reserve. A classification of bones based on shape will give some help in determining which bones contribute to the function of blood formation.

Long bones are longer than they are wide, and are relatively ovoid in cross section. These include the main bones of the arms and legs, the ribs, and several others. Most blood formation takes place in the ends of long bones, though some other types also contribute.

Short bones are rather cubical in shape; there is usually little difference in length, width, and thickness. The bones of the wrist and ankle are examples of short bones.

Flat bones are not necessarily perfectly flat, but they are relatively thin, and constant in cross section. The bones of the skull are good examples of flat bones. Some blood formation also takes place within these bones.

Irregular bones are of complex shapes and do not fit into any of the other categories. They contribute all four of the functions of the bone.

Read the descriptions of bones below. After each, write the type of bone and indicate which functions of the skeletal system it seems to fill.

(a) The scapula, or shoulder blade, is roughly triangular, with a thin cross section in most of the area. It is located on the back, between ribs and skin, and many arm muscles are attached to it.

Type ______________ Function ______________

(b) The humerus, or upper arm bone, extends from the shoulder to the elbow. It has an active marrow cavity at each end.

Type ____________________ Function ____________________

(c) The capitate, a wrist bone, is small, roughly cubical, and as thick as it is wide.

Type ____________________ Function ____________________

(d) The frontal bone, which forms the forehead, is large in surface area, but quite thin. It contains some marrow, and is an attachment for many muscles of the head.

Type ____________________ Function ____________________

(e) The vertebrae, or bones of the spinal column, have complex shapes. They are wider than they are long, and are not thin in cross section. Many projections from the vertebrae are attachments for muscles. Each vertebra has a central opening through which the spinal cord passes, making a bony tube down the center of the back. A marrow cavity is located in the body of each vertebra.

Type ____________________ Function ____________________

- - - - - - - - - - - - - - - - - -

(a) flat bone; mineral reserve, support, and blood formation
(b) long bone; mineral reserve, support, and blood formation
(c) short bone; mineral reserve, support
(d) flat bone; mineral reserve, support, protection, and blood formation
(e) irregular bone; mineral reserve, support, protection, blood formation

(Notice that these bones are all involved in mineral reserve and support. This is generally true of all bones.)

3. Regardless of the functions or shapes of bones, all are formed and grow in the same general way. Bone-forming cells called osteoblasts (osteo means bone) produce a substance which is then hardened by calcium, or calcified. Some osteoblasts become trapped in the hardening substance and converted into osteocytes which continue to live within the bone itself to nourish it.

Bone is thus a living tissue, which explains why a break can heal, or a person can grow. As children grow, their bones are in a constant state of change, accommodating other growing structures and remodeling themselves to meet the demands of developing muscles and such other structures as teeth. The substance produced by the osteoblasts is called osteoid. The outer covering of the bone is thus called periosteum (peri means around).

The lining of a marrow cavity is called endosteum; it is completely located inside the bone. Some other terms are also useful to know before we begin examining the body bones. Long bones have a shaft and two ends. The shaft is called the diaphysis, while the ends are called epiphyses. Most growth in length takes place at the epiphyseal line, which joins the shaft to each end. Growth in width of long bones takes place just beneath the periosteum.

(a) Specify whether each bracket in the figure here indicates an epiphysis or diaphysis. How would this bone be classified?

____

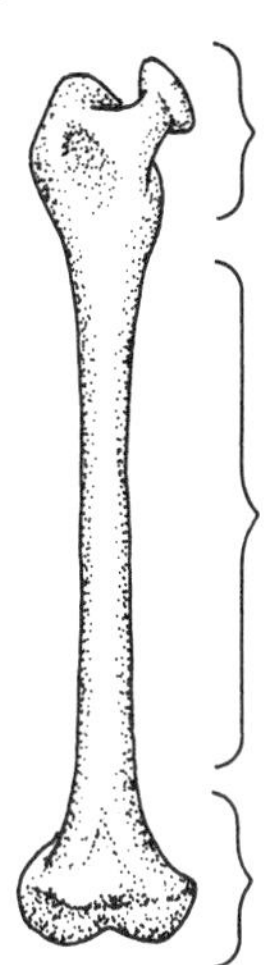

(b) This figure is a cross section of the bone in (a) above. Label the following on the diagram: endosteum, periosteum, marrow cavity, and bony portion.

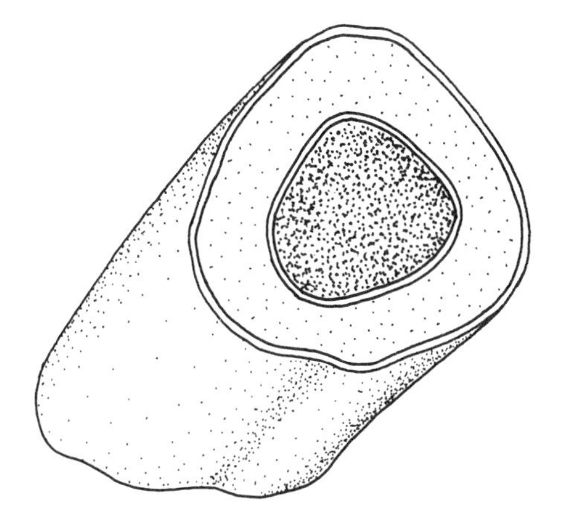

(c) What type of cell produces bones? ____________________

(d) What type of cell lives within already formed bones? __________

____

(e) What cell produces osteoid? ____________________ What does osteoid become? ____________________

\- - - - - - - - - - - - - - - - - -

(a) Each end is an epiphysis; the shaft is diaphysis. This is a long bone.
(b) (See figure at right.)
(c) osteoblast
(d) osteocyte
(e) It is produced by osteoblasts, then calcified into bone.

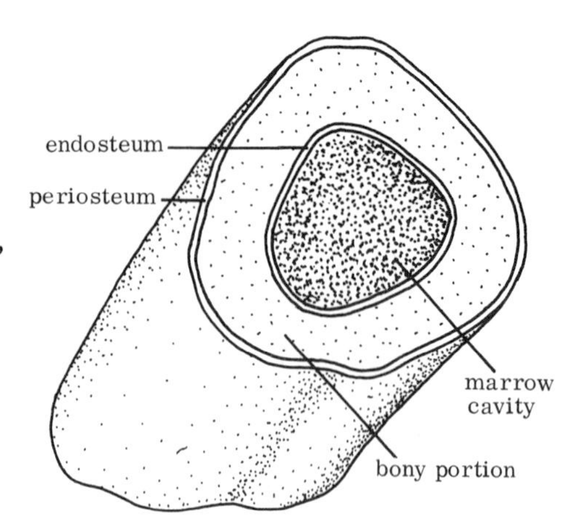

4. In the human skeleton, all the bones except one are continuous (adjacent). For ease of study, the skeleton can logically be divided into two parts from a functional point of view. The appendicular skeleton includes the bones of the arms and legs with the corresponding shoulder and hip bones. In biology, an appendage is any part that is joined to a trunk, or axis, hence the term appendicular skeleton. The axial skeleton, in turn, includes the skull, backbone, and the bones of the thorax, as well as the single disconnected hyoid bone in the neck.

Write whether the structures indicated in the diagram at right are part of the axial or appendicular skeleton.

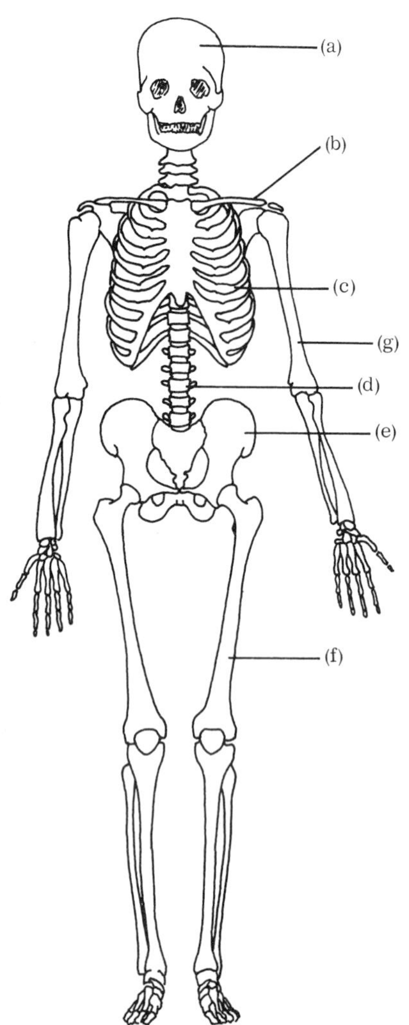

(a) ____________________

(b) ____________________

(c) ____________________

(d) ____________________

(e) ____________________

(f) ____________________

(g) ____________________

- - - - - - - - - - - - - - - - - - - -

5. (a) axial; (b) appendicular; (c) axial; (d) axial; (e) appendicular; (f) appendicular; (g) appendicular

THE AXIAL SKELETON

5. The axial skeleton includes 80 bones, which at first glance seems like a lot to learn. However, 24 of these are ribs and 24 are vertebrae, which quickly cuts down the number to be studied. The skull presents most of the problems in studying the axial skeleton. It can be separated into cranial bones, which encase the brain, and facial bones, which make up the rest of the head. Following is a breakdown of bones in the axial skeleton.

<u>Head</u>		
Cranium	8 bones	
Face	14 bones	
Ear	6 bones	
Hyoid	1 bone	
		29
<u>Vertebral column</u>		
Vertebrae	24	
Sacrum	1	
Coccyx	1	
		26
<u>Thorax</u>		
Ribs	24	
Sternum	1	
		25
	total	80 bones

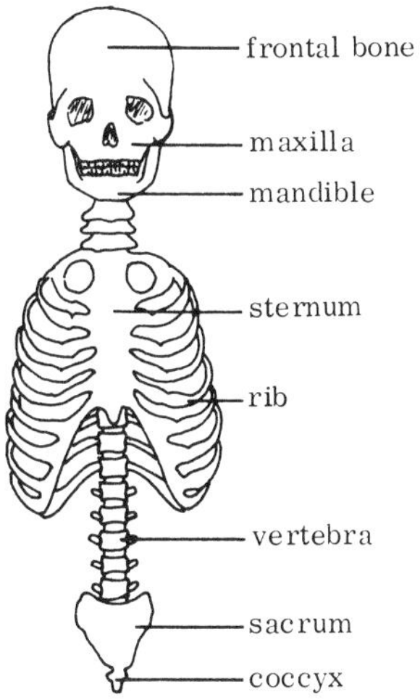

In the diagram above, many of the bones of the axial skeleton are labeled.

(a) Is the frontal bone part of the cranium or of the face? ____________

____

(b) Name the bone that makes up the lower jaw? ____________________

(c) The bony thorax actually is made up of 37 bones. Can you tell why the listing above specified only 25? ____________________

____

(d) What bone in the vertebral column is widest? ____________________

- - - - - - - - - - - - - - - - - -

(a) cranium (it helps encase the brain); (b) mandible; (c) 12 vertebrae are listed as part of the spinal column; (d) sacrum

Cranial and facial bones

6. The cranium is made up of eight bones. Of these, two are paired (the parietals and temporals) and found on right and left sides, while the remaining four are centered in the midline. For this reason, only six bones are labeled in the figure below. All the bones are evident from the inside and the outside of the skull. But one, the ethmoid, is only seen as part of the orbit (eye socket) from the outside.

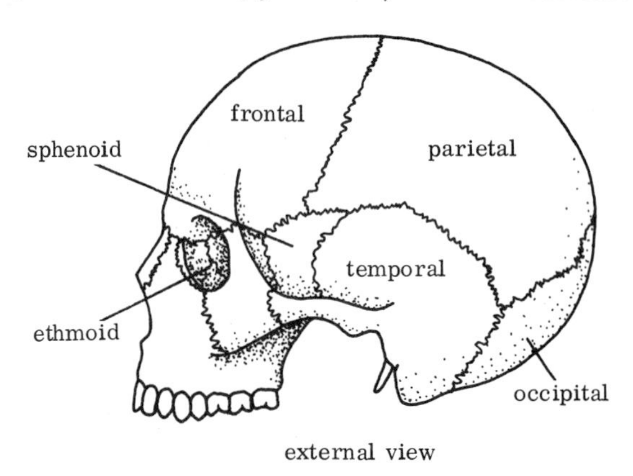

external view

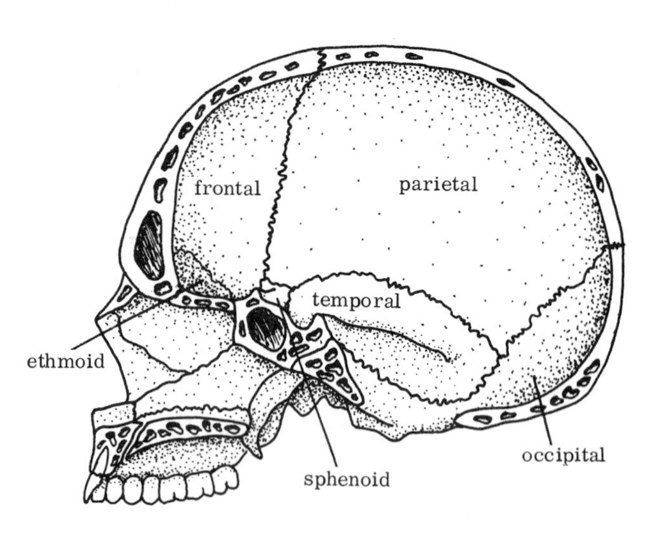

median view

(a) Five bones are joined to the right parietal bone by sutures, or fixed joints. Can you name them? ______________________

____

____

(b) What bone makes up part of the internal nasal septum (dividing the nasal cavity into two parts), as well as forming part of the brain?

____

(c) On the back of your head just above the neck you can feel a bump. On what bone is this protuberance located? _______________

(d) How are the frontal and parietal bones classified in terms of shape?

____

(e) What is the major function of the eight cranial bones? _______________

____

- - - - - - - - - - - - - - - - - -

(a) left parietal, frontal, occipital, right temporal, and right sphenoid
(b) ethmoid
(c) occipital
(d) flat bones
(e) protection of the brain

7. The occipital bone has a large opening (just anterior to the bump you felt) through which the spinal cord passes. This is called the foramen magnum (foramen means opening; magnum means large). The occipital bone, like the parietal and frontal, is basically a flat bone.

The temporal bones house the hearing and balance senses of the body. Within each temporal bone is the middle ear cavity, which contains three small bones called ossicles. We shall examine the function of these ossicles in more detail in Chapter 6. The temporal bone has a flat area, but is mainly irregular, as are the tiny ossicles.

The sphenoid and ethmoid are very irregular in shape. They also contribute to the facial structures since parts of both help form the orbit or eye socket as well as many internal structures.

(a) Through which bone does the spinal cord pass? _______________ What is the opening called? _______________

(b) In which bone are three other bones encased? _______________ What is the cavity called? _______________ What are the small bones called? _______________

(c) Which of the cranial bones are classified as irregular? _______________

____

(d) Can you list the eight cranial bones?

Single bones: ________________ Paired bones: ________________

________________ ________________

____

____

- - - - - - - - - - - - - - - - - -

(a) occipital; foramen magnum
(b) temporal; middle ear cavity; ossicles
(c) ethmoid, sphenoid, right temporal, and left temporal
(d) Single bones: frontal, occipital, sphenoid, ethnoid;
Paired: parietal, temporal

8. The rest of the skull is made up of the 14 facial bones. Twelve of these are in six pairs while only two are centered in the midline. The following diagram shows 9 of the 14 bones. The mandible, or lower jaw, is the only skull bone that can be moved separately from the skull, allowing for speech and chewing. The mandible forms a joint, or articulates, with the temporal bones on each side at the temporomandibular joints. All other skull bones are joined to each other by fixed joints called sutures.

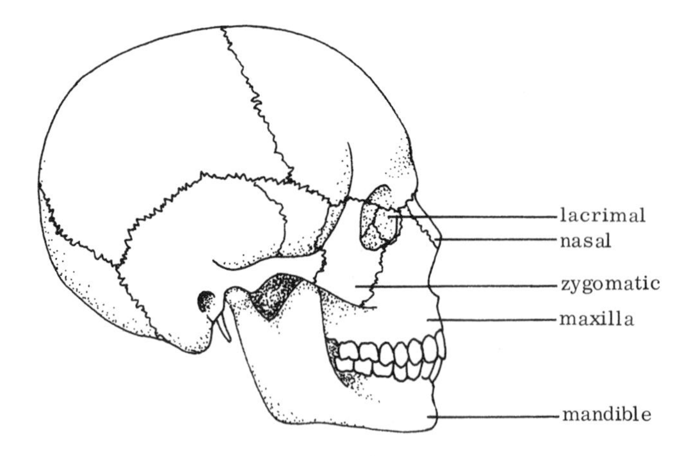

The maxillae (singular, maxilla) make up the upper jaw, which also has teeth. The zygomatic bone joins to a process or projection of the temporal bone to form the zygomatic arch or cheekbone. The nasal bones are located at the sides of the top of the nose. The lacrimal bones are very small and make up a small part of the orbit.

The facial bones not shown in the figure are all located in the nasal cavity and do not extend to the surface. The two palatine bones are located behind the maxillae, while the two inferior nasal conchae are located on the posterior lateral walls of the nasal cavity. The midline-located vomer extends from the front of the nasal septum backward to the sphenoid bone.

(a) Which facial bones make up the upper and lower jaws? ____________

____

(b) All 14 facial bones are in the same shape classification. What type of bones are they? ____________________

(c) Name the facial bones that make up the nasal cavity walls and septum.

Single bone: ______________ Paired bones: ____________________

____

____

____

(d) Can you name the bones that form the wall of each orbit?

Cranial: ____________________ Facial: ____________________

____________________ ____________________

____________________ ____________________

(e) The cheekbone is made up of processes from one cranial and one facial bone. What are they?

Cranial: ____________________ Facial: ____________________

(f) What is the only movable joint in the skull? ____________________

What are the nonmovable joints called? ____________________

- - - - - - - - - - - - - - - - - -

(a) two maxillae, mandible
(b) irregular bones
(c) Single bone: vomer; Paired bones: maxillae, palatines, inferior nasal conchae, nasals
(d) Cranial: frontal, ethmoid, sphenoid; Facial: lacrimal, maxilla, zygomatic
(e) Cranial: temporal; Facial: zygomatic
(f) temporomandibular joint; sutures

9. (a) The cranial and facial bones make up the skull. Name the bones centered on the midline.

Cranial: ____________________ Facial: ____________________

____________________ ____________________

____

____

(b) Name the paired cranial and facial bones described below.

Top of skull ____________________

Contain middle ear ossicles ____________________

Upper jaw ____________________

Facial part of cheekbone ____________________

Sides of top of nose ____________________

Small bone in orbit ____________________

Back part of palate ____________________

Small bone in lateral nasal cavity ____________________

(c) What two cranial bones contribute greatly to the internal facial structure as well? ____________________

- - - - - - - - - - - - - - - - - -

(a) Cranial: frontal, occipital, sphenoid, ethmoid; Facial: mandible, vomer

(b) parietals; temporals; maxillae; zygomatic; nasal; lacrimal; palatine; inferior nasal conchae

(c) sphenoid and ethmoid

10. The 29 bones of the skull include 8 cranial and 14 facial bones. In addition, 6 ossicles are located within the temporal bones. The single hyoid bone is located in the anterior upper neck. This U-shaped bone is attached to muscles of the tongue and neck.

(a) Consider the cranial bones. How many are paired? ____________

How many singles? ____________________

(b) Consider the facial bones. How many are paired? ____________

How many are single? ____________________

(c) Consider the other skull bones. How many are paired? __________

How many single? ____________________

(d) List the bones of the skull.

Cranial	Facial	Other
____________	____________	____________
____________	____________	____________
____________	____________	
____________	____________	
____________	____________	
____________	____________	
	____________	
	____________	

- - - - - - - - - - - - - - - - - - -

(a) 2, 4
(b) 6, 2
(c) 3, 1
(d)

temporal (2)	maxillae (2)	ossicles (6)
parietal (2)	palatine (2)	hyoid
frontal	zygomatic (2)	
occipital	nasal (2)	
ethmoid	inferior nasal conchae (2)	
sphenoid	lacrimal (2)	
	vomer	
	mandible	

Spinal and thoracic bones

11. The spinal cord begins where it descends through the foramen magnum of the occipital bone. Throughout its length, the spinal cord is housed in a canal down the middle of the vertebral column, or spine.

The vertebral column is made up of 24 separate vertebrae, a sacrum, and a coccyx. This strong, flexible column supports the body and protects the spinal cord. The first 7 vertebrae are the cervical vertebrae (cervical means neck). The next 12 are thoracic vertebrae; ribs are attached to these vertebrae. Five lumbar vertebrae make up the "small of the back." The single sacrum is made up of 5 fused vertebrae; two large hip bones are attached to the sacrum. The coccyx (tailbone) is also composed of about 5 fused bones. These fused bones are separate in childhood, but fuse together to form the sacrum and coccyx in the adult.

(a) Which of the four functions of bone are most important in vertebrae?

____

(b) Name the groups of vertebrae indicated on the right.

A ____________________

B ____________________

C ____________________

D ____________________

E ____________________

(c) With what bone would you expect the superior portion of the vertebral column to articulate? ____________

____

(d) Other bones articulate with 13 vertebrae. Which vertebrae and which bones? ____________________

____

- - - - - - - - - - - - - - - - - -

(a) support and protection (they also contribute to blood formation and the mineral reserve)
(b) A—cervical (7); B—thoracic (12); C—lumbar (5); D—sacrum; E—coccyx
(c) occipital bone (near foramen magnum)
(d) 12 thoracic vertebrae and ribs; sacrum and hip bones

12. While vertebrae in different locations show variations, the basic plan is very much the same. The body of each vertebra is toward the anterior

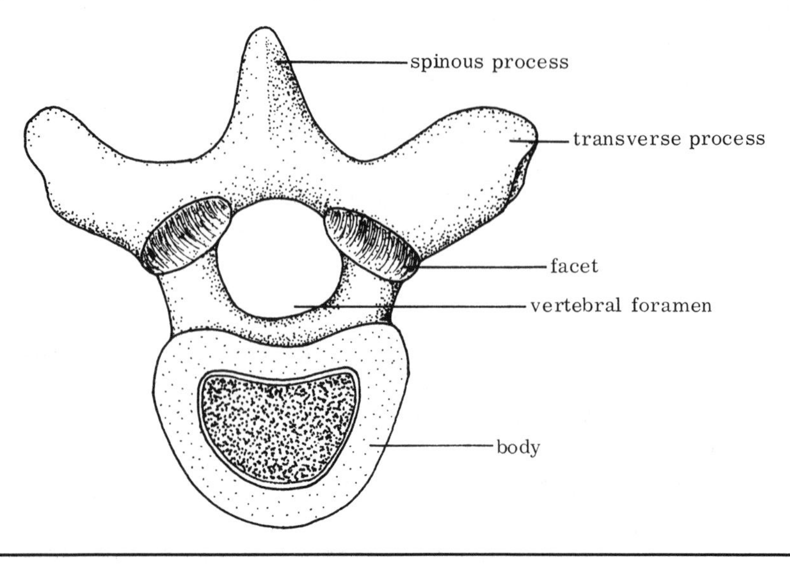

and contains a marrow cavity. A vertebral arch completes a ring enclosing the vertebral foramen. A spinous process (sharp projection) extends posteriorly, while transverse processes extend laterally. In addition, thoracic vertebrae have facets for attaching ribs, while the lumbar vertebrae are significantly more bulky than are the cervicals.

The first two cervical vertebrae are extensively modified to form the highly mobile articulation with the skull. In fact, they may not look like vertebrae to untutored eyes.

The sacrum is roughly triangular and obviously a single bone. The coccyx, while also triangular, is much smaller and the original bones, while fused, can be easily counted.

(a) Feel your spine. What part of a vertebra causes the distinct bumps?

____

(b) What is found in the vertebral foramen of a living body? __________

____

(c) What process takes place in the body of a vertebra? __________

____

(d) How many vertebrae would you expect to be similar to the drawing in this frame? ______________________________

- - - - - - - - - - - - - - - - - -

(a) spinous process; (b) spinal cord; (c) blood formation; (d) 22 (all except top two cervical)

13. The bones of the thorax include the 12 thoracic vertebrae, 24 ribs, and one sternum, or breastbone, in the anterior midline. The ribs are attached to facets on the thoracic vertebrae, one on each side of each vertebra. In the front, the sternum serves as a point of attachment for most of the ribs by means of costal cartilages (costa means rib).

The upper seven ribs on each side are called true ribs as their costal cartilages are directly attached to the sternum. The next three are the false ribs, because their costal cartilages are attached to cartilages of the seventh rib. The last two are floating; their anterior ends are not attached to cartilage or bone.

The curving ribs are sometimes considered to be flat bones because of their cross sectional shape, and are sometimes considered to be long bones. They have a head on the posterior end for attachment to vertebrae.

The flat sternum in the midline has three parts. The elongated body which comprises most of its length is the site of attachment for most of the costal cartilages. The manubrium is the superior portion; it is the point of articulation of the axial skeleton with the upper appendicular skeleton, as well as the site of the first costal cartilage. You can feel a midline notch in the manubrium at the anterior base of your neck. The xiphoid process at the lower end of the sternum is triangular and, at its junction with the body, receives the seventh costal cartilage.

(a) Costal cartilages from how many ribs terminate at the junction of the xiphoid process and the body of the sternum? ______________

____

(b) Indicate whether each of the following is a true rib, a false rib, or a floating rib.

rib 1 ______________

rib 7 ______________

rib 9 ______________

rib 11 ______________

(c) Of the bony thorax, how many bones could be classified as flat?

____

(d) Name the part of the sternum that articulates with the appendicular skeleton. ______________

(e) Describe the attachments of both ends of one eighth rib.

anterior ______________

posterior ______________

- - - - - - - - - - - - - - - - -

(a) two—right and left seventh ribs
(b) true rib, true rib, false rib, floating rib
(c) 25 (24 ribs and sternum)
(d) manubrium
(e) anterior: costal cartilage of eighth rib is attached to costal cartilage of seventh, which is attached to sternum; posterior: head of rib is attached to facet on eighth thoracic vertebra

THE APPENDICULAR SKELETON

14. The second division of the skeleton includes the bones of the appendages (arms and legs) with their attachments to the axial skeleton. The shoulder region, called the pectoral girdle, includes two bones on each side. The hip region, called the pelvic girdle, has just one large bone on each side. The pectoral girdle has its attachment to the axial

skeleton at the superior aspect of the sternum. The pelvic girdle is joined to the axial skeleton at the sacrum, forming the sacroiliac joint.

(a) How do the pelvic and pectoral girdles differ in number of bones?

____

(b) Is the pectoral girdle joined to the axial skeleton in the anterior or posterior region? ____________

(c) Is the pelvic girdle joined to the axial skeleton in the anterior or posterior region? ____________

\- - - - - - - - - - - - - - - - - -

(a) pelvic has one bone on each side, pectoral two
(b) anterior (at sternum)
(c) posterior (at sacrum)

15. The bones of the pectoral girdle are the clavicle (collar bone) and scapula (shoulder blade). The clavicle is joined to the axial skeleton at the sternoclavicular joint (A in the following diagram). The distal end of the clavicle articulates with the acromion process of the scapula, forming the acromioclavicular joint (B in the diagram).

While the clavicle is a long narrow bone, the scapula is roughly triangular with the base superior, and the apex below. On the lateral surface of the scapula is the glenoid fossa, a depression for articulation with the bone of the upper arm. A long spine on the posterior scapular surface leads into the projection called the acromion process, which articulates with the clavicle.

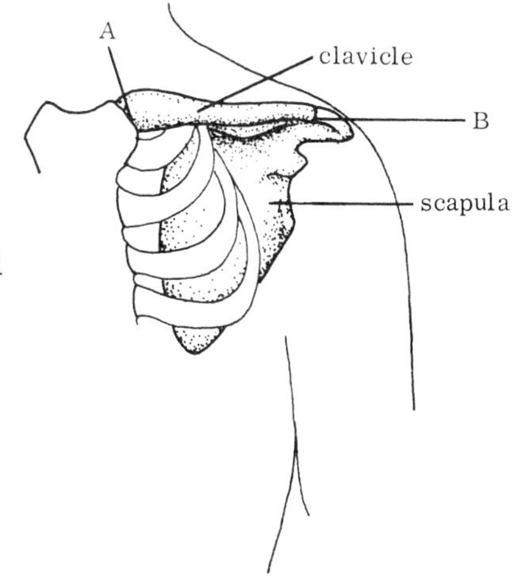

(a) Which of the bones of the pectoral girdle attaches it to the axial skeleton? ____________

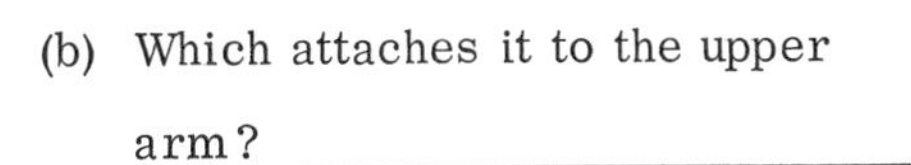

(b) Which attaches it to the upper arm? ____________

(c) Name the bones shown below.

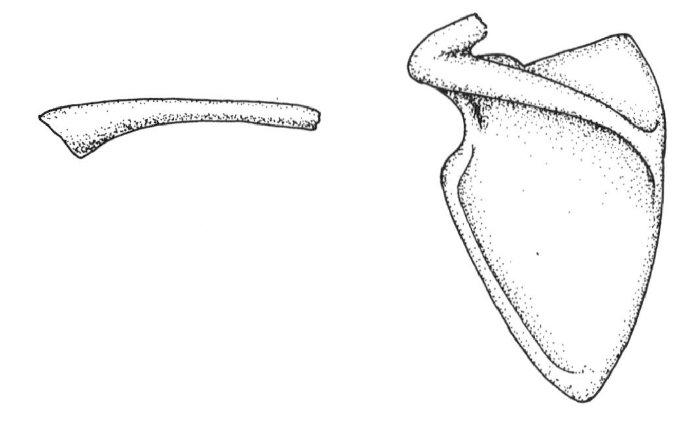

(1) ________________ (2) ________________

(d) Which point on the clavicle indicated in the diagram articulates with the scapula? ____________________

(e) Name the joint described in (d). ____________________

- - - - - - - - - - - - - - - - - -

(a) clavicle
(b) scapula (glenoid fossa)
(c) (1) clavicle, (2) scapula
(d) B in the diagram
(e) acromioclavicular joint

16. The arm consists of three bones: the humerus in the upper arm, and the radius and ulna in the forearm. The elbow is formed by a projection on the end of the ulna.

The humerus articulates with the glenoid fossa of the scapula. This long bone has a head at the proximal end which fits into the glenoid fossa. The shaft is fairly smooth and straight. The distal end of the humerus has fossae (a fossa is just a depression in a bone; it usually is part of a joint) for joints with both the ulna and radius, called olecranon and radial fossae.

The olecranon process of the ulna forms the tip of the elbow. It curves around to glide smoothly over the distal humerus as the joint moves. The distal ulna is shaped to articulate, in conjunction with the radius, with the bones of the wrist. The radius is located lateral to the ulna in the anatomical position. At both ends, the radius articulates with the ulna. Its shaft is relatively uniform in thickness.

(a) In the forearm of an individual, which of the bones is longer?

____

(b) The humerus articulates with three other bones. Name the bones, and the portion of the humerus that connects with each. __________

____

____

(c) What structure forms the elbow? ________________ What bone is this an extension of? ________________

(d) You may occasionally bump your elbow, and get a tingling sensation; you have hit your "funny bone." Can you tell from the information above why that term is used? ________________________

- - - - - - - - - - - - - - - - - -

(a) ulna
(b) scapula with the head; ulna with the olecranon fossa; radius with the radial fossa
(c) olecranon process of ulna
(d) It's at the end of the humerus ("humorous") bone. (Actually, of course, it's at the end of the ulna, but that doesn't help.)

17. The pectoral girdle contains 2 bones. The arm contains 3 more. The wrist adds another 8 while the hand contains 19 bones.

The bones of the wrist are called carpal bones. These eight, almost cuboidal, bones are roughly arranged in two rows of four. The joints between them move only a little. The metacarpal bones make up the palm area of the hand, while the phalanges make up the fingers, with a joint at each knuckle. A metacarpal extends from the wrist to each finger, thus there are five metacarpals.

As you can tell from examining your hand, each thumb contains two bones, or phalanges, a proximal phalanx that articulates with the metacarpal, and a distal phalanx. Each other finger has three phalanges. While the phalanges are rather short in length, they are classified as long bones. The carpals are classified as short bones. The upper arm, forearm, wrist, and hand are known together as the upper extremity; corresponding parts in the leg and foot make up the lower extremity.

(a) How many bones make up the upper extremity?

upper arm __________ wrist __________

forearm __________ hand __________

total __________

(b) Which point (A, B, or C) in the diagram indicates the metacarpal-phalangeal joint of the thumb?

____

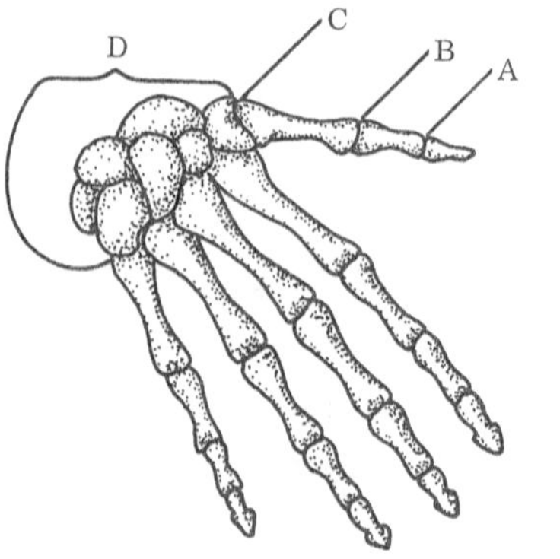

(c) Name the bones indicated by D.

____

(d) How many other bones articulate with bones of the wrist?

____

(e) How many long bones are in the upper extremity? __________

How many short bones? __________

- - - - - - - - - - - - - - - - - -

(a) 1, 2, 8, 19 = 30 bones per arm
(b) B
(c) carpals
(d) 7 (radius, ulna, and 5 metacarpals)
(e) 22; 8

18. (a) The bones shown here are all about the same size. Name them.

A ______________________

B ______________________

C ______________________

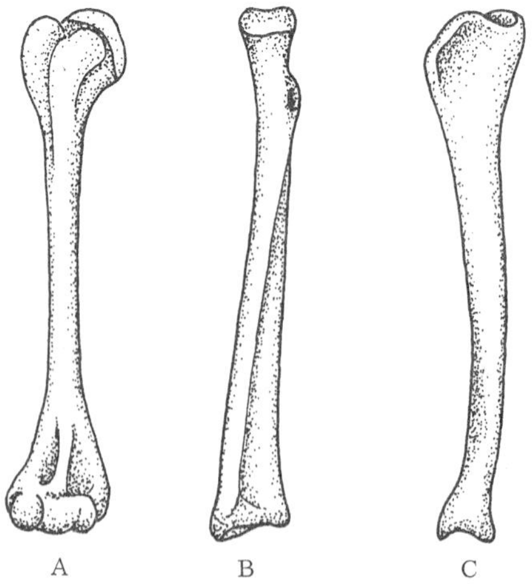

(b) On the humerus, label the point of articulation with the scapula.

(c) On the ulna, label the olecranon process.

(d) On the radius, label the distal end.

(e) On the radius, label the shaft.

(f) Is a carpal or a metacarpal more similar to a humerus? Why?

____

____

(g) How many bones are located between the tip of your little finger and the wrist? ______________________________

- - - - - - - - - - - - - - - - - - -

(a)-(e) (See figure at right.)
(f) metacarpal, because it has two ends and a shaft, and both are long bones
(g) four (three phalanges and a metacarpal)

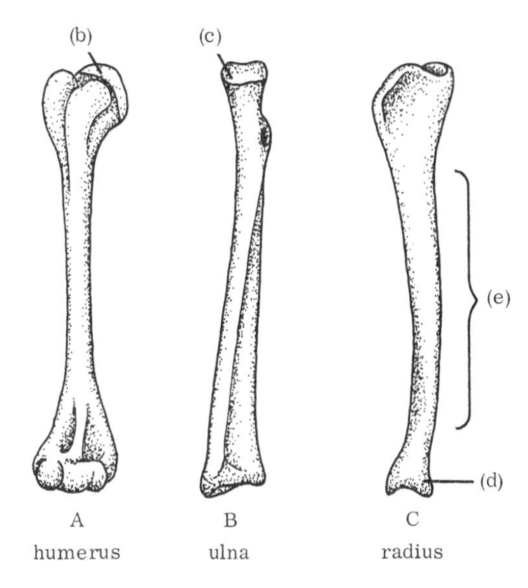

19. The pelvic girdle, or hip, contains one bone on each side. The two os coxae are firmly connected to the sacrum posteriorly and join in the anterior midline at the symphysis pubis. In a newborn child, each os coxa is made up of three bones. These bony areas can still be identified in the adult bone, although they have long since fused, becoming physically, as well as functionally, one bone.

The large superior portion (A in the diagram) is the ilium. Its edge is called the iliac crest. The ischium (B) is the inferior posterior part of the os coxa. The pubis (C) is the anterior inferior part, which joins with its opposite number at the symphysis.

All three portions contribute to forming the walls of the acetabulum,

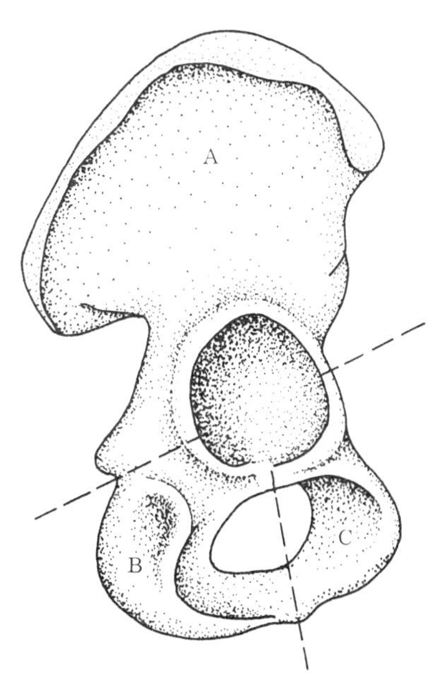

a large depression for the joint with the leg. Another distinctive feature is the obturator foramen, the opening in the lower portion of each os coxa.

(a) Name the pair of bones that makes up the pelvic girdle. ____________

____

(b) To what is the pelvic girdle attached posteriorly? ________________

(c) What two parts of the os coxa make up the edges of the obturator foramen? ______________________

(d) What do you call the long upper edge of the pelvic bone? __________

____

(e) The three portions of the pelvic bone join in a depression that articulates with the leg. This depression is called the _______________

- - - - - - - - - - - - - - - - - -

(a) os coxae; (b) sacrum; (c) ischium and pubis; (d) iliac crest; (e) acetabulum

20. The shape of the os coxae differ in males and females, as an adaptation for pregnancy and child bearing. The iliac portions flares out more in the female, resulting in broader hips. The pelvis is thus shallower but wider. The anterior-posterior dimension of the base of the pelvis is wider in the female, producing a rounded opening as opposed to the heart-shaped opening in the male.

(a) Label these pelvic girdles as male or female.

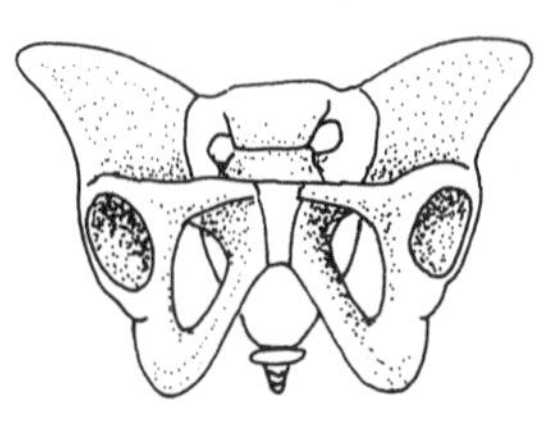
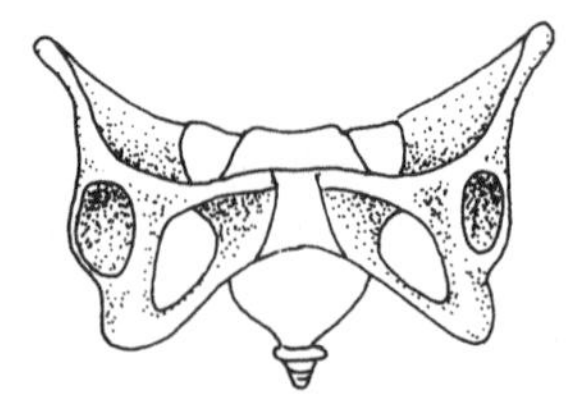

(b) Label these pelvic girdles as male or female.

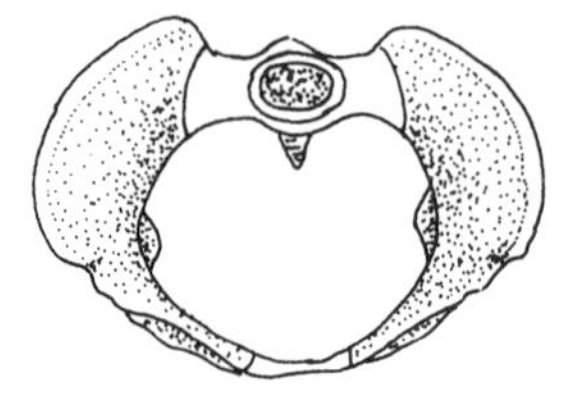
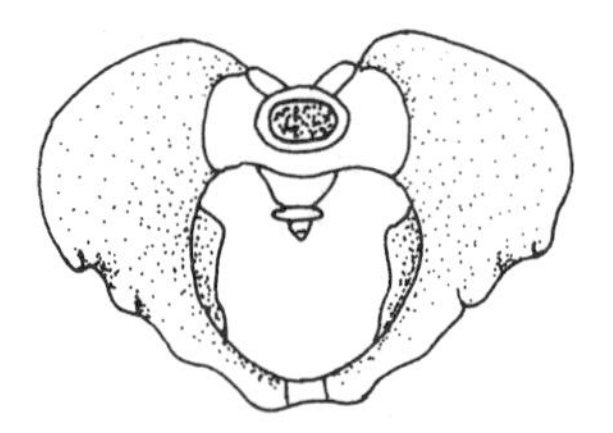

(c) Name the bone shown at the right. ________________
Identify its parts.

A ____________________

B ____________________

C ____________________

D ____________________

E ____________________

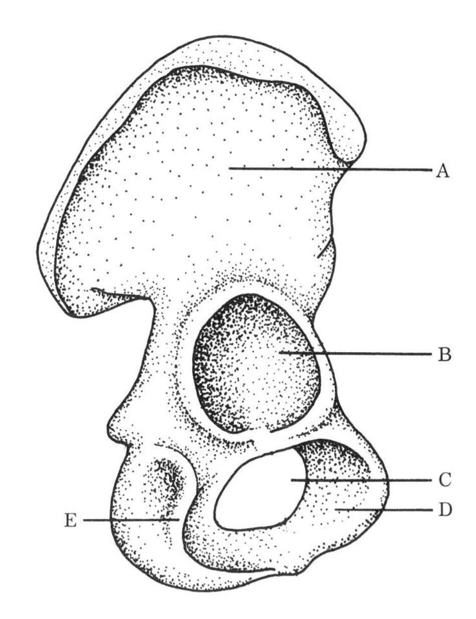

- - - - - - - - - - - - - - - - - -

(a) male on the left, female on the right
(b) female on the left, male on the right
(c) os coxa; A—ilium, B—acetabulum, C—obturator foramen, D—pubis, E—ischium

21. The number and distribution of the bones of the leg are essentially similar to those of the arm. One bone in the upper leg, the femur, corresponds to the humerus in the arm. The proximal end of the femur is capped by a large head, extending medially from the femoral shaft and resting in the acetabulum. The distal end of the femur is expanded and includes two condyles (medial and lateral) for articulation with the tibia of the lower leg. (Condyles are rounded projections on bones.)

One extra bone, the patella or knee cap, also articulates with condyles of the distal femur. The patella is an irregular bone that is embedded in the tendons of the knee.

(a) To which bone of the upper extremity does the femur correspond?

____

(b) List the femoral structures that correspond with:

radial and olecranon fossae ____________________

head of humerus ____________________

glenoid fossa ____________________

(c) Name three bones that articulate with the femur. ________________

____

- - - - - - - - - - - - - - - - - - -

(a) humerus
(b) (medial and lateral) condyles; head of femur; acetabulum
(c) os coxa, tibia, and patella

22. The tibia and fibula form the skeleton of the lower leg. The tibia is longer and thicker, and articulates with the distal femur. Like the distal femur, the proximal tibia has a lateral and medial condyle for articulation.

The distal tibia has a curved extension called the medial malleolus that articulates with one of the ankle bones. You can easily feel the medial malleolus in your ankles—it is the bump on the medial side. The bump on the lateral side is the lateral malleolus—a distal extension of the fibula, which also articulates with the ankle.

The fibula is slender, and its proximal end articulates with the tibia just below the lateral condyle. The distal end also forms a joint with the tibia, as well as with the ankle.

(a) Indicate which of the lower leg bones is more similar to:

radius __________________________

ulna __________________________

(b) On the ankle, you can feel the lateral and medial malleoli. Name the bone of which each is part.

lateral malleolus __________________________

medial malleolus __________________________

(c) Name three bones that articulate with the tibia. ____________________

____

(d) How many bones articulate with the fibula? ____________________

(e) Name the projections on the tibia that articulate with the femur.

____

(f) Name the structure that forms the tip of the elbow. ________________

Of the knee. __________________________

- - - - - - - - - - - - - - - - - - -

(a) fibula; tibia
(b) fibula; tibia
(c) femur, fibula, and ankle bone
(d) 2 (tibia in two places, ankle bone)
(e) medial and lateral condyles
(f) olecranon process; patella

23. (a) The bones on the right are all drawn the same size. Can you name them by examining their projections?

A ____________________

B ____________________

C ____________________

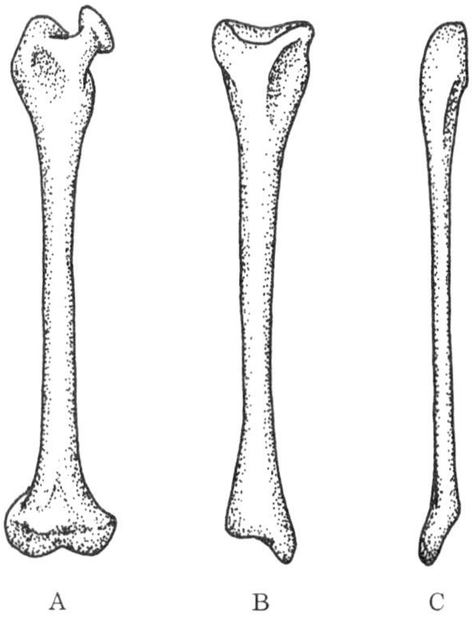

A B C

Indicate the long leg bone with which each of the following is associated:

(b) head (of bone) ____________________

(c) lateral malleolus ____________________

(d) medial and lateral condyles ____________________

(e) medial malleolus ____________________

(f) patella ____________________

- - - - - - - - - - - - - - - - - -

(a) A—femur; B—tibia; C—fibula
(b) femur
(c) fibula
(d) femur and tibia
(e) tibia
(f) femur (only articulation with patella)

24. Structurally, the foot is very similar to the hand. The only variations are seen in the ankle bones or tarsals, of which there are seven. Of these tarsals, the heel bone, calcaneus, is very large, while the talus

bone, which articulates with both the tibia and fibula, bears the entire weight of the body. The other five tarsal bones are smaller and more like the carpals in the wrist.

The number and distribution of metatarsals is the same as in metacarpals, while the quantity of phalanges is also the same. The first (great) toe, like the thumb, has only two phalanges while each other digit contains three. The phalanges in the toes are generally shorter than in the fingers, and the joints formed by them are less movable or flexible.

(a) Fill in the comparison chart below.

	Upper extremity	Lower extremity
Bone name (wrist or ankle)	______	______
Number in one extremity	______	______
Bones in flat of hand or foot (name)	______	______
Number	______	______
Bones in digits (name)	______	______
Number	______	______
Total number of bones	______	______

(b) Name the two largest ankle bones, and indicate why they are different from the other tarsals.

____

____

- - - - - - - - - - - - - - - - - -

(a)

carpals	tarsals
8	7
metacarpals	metatarsals
5	5
phalanges	phalanges
14	14
27	26

(b) calcaneus—forms heel
talus—takes weight of body

Summary: Appendicular skeleton

25. (a) What bone is roughly triangular, has a posterior spine, and a glenoid fossa? ______________

(b) In what bone do you find an obdurator foramen? ______________

(c) What bones are joined by the acromioclavicular joint? __________
____

(d) Name the three bones that articulate at the ankle. ____________
____

(e) Label the bones in the diagram at right.

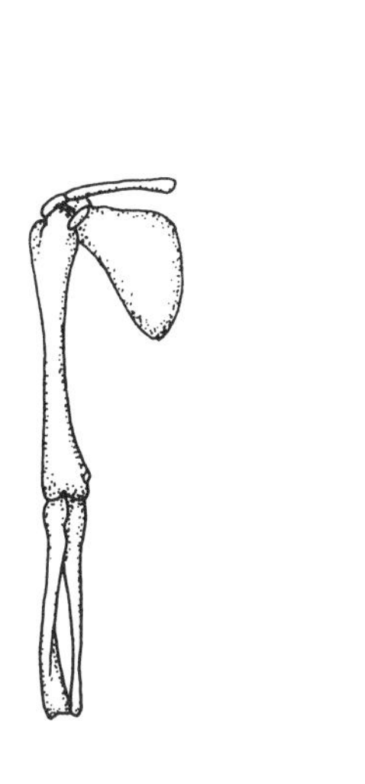

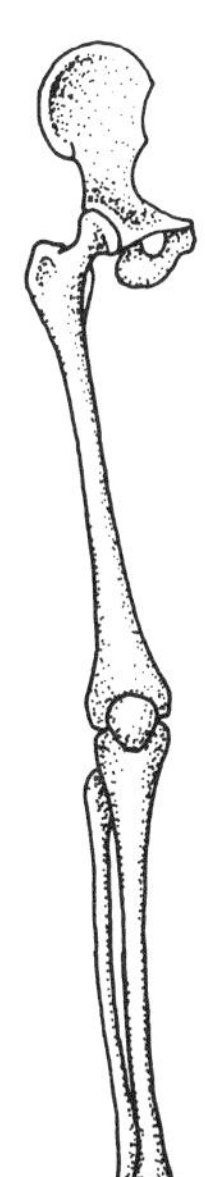

- - - - - - - - - - - - - - - - - -

(a) scapula
(b) os coxa(e)
(c) scapula and clavicle
(d) tibia, fibula, and talus
(e) (See figure at right.)

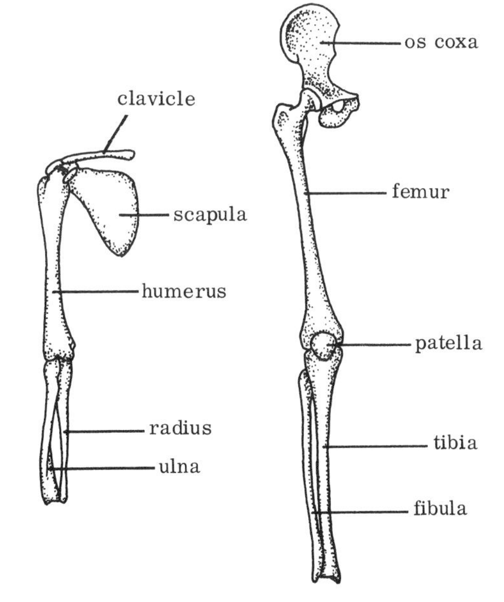

SELF-TEST

This Self-Test is designed to show how well you have mastered this chapter's objectives. Answer each question to the best of your ability. Correct answers and review instructions are given at the end of the test.

1. Which functions of the skeletal system are met by each of the bones listed below?

   (a) humerus ____________________

   (b) sternum ____________________

   (c) scapula ____________________

2. In the formation of bones, what events causes some osteoblasts to be converted into osteocytes? ____________________________________

3. Classify the bones listed below according to shape.

   (a) phalanx____________________

   (b) wrist bone ____________________

   (c) parietal ____________________

   (d) metatarsal ____________________

   (e) vertebra ____________________

4. Identify the bones listed below as being in the axial or appendicular skeleton.

   (a) os coxa ____________________

   (b) thoracic vertebrae ______________

   (c) lacrimals ____________________

   (d) scapulae ____________________

5. Name the bones indicated in the diagram on the following page.

   A ____________________

   B ____________________

   C ____________________

   D ____________________

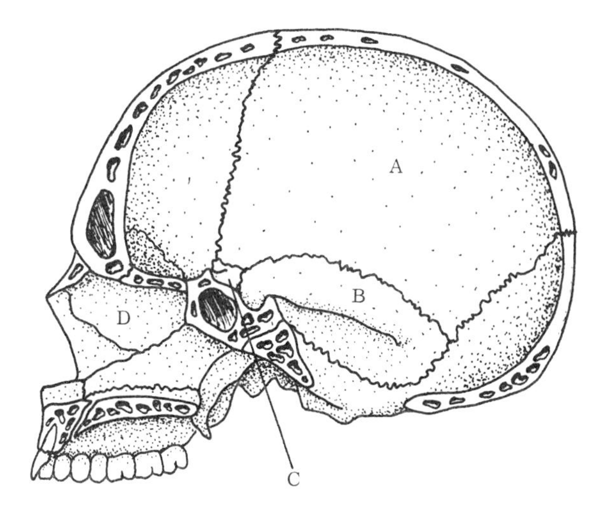

6. Name the unpaired bones of the cranium and face.

   Cranium: ____________________ Face: ____________________

   ____________________ ____________________

   ____

   ____

7. Name the bone(s) on which each of the following is found.

   (a) foramen magnum ____________________

   (b) acetabulum ____________________

   (c) acromion ____________________

   (d) medial malleolus ____________________

   (e) olecranon process ____________________

   (f) manubrium ____________________

   (g) costal cartilage ____________________

8. List the names of the bones in the lower appendicular skeleton. (The number of each type of bone in one extremity is given.)

   ____________________ (1) ____________________ (1)

   ____________________ (1) ____________________ (7)

   ____________________ (1) ____________________ (5)

   ____________________ (1) ____________________ (14)

Answers

Compare your answers to the Self-Test questions with those answers given below. If all of your answers are correct, you are ready to go on to the next chapter. If you missed any, review the frame indicated in parentheses following the answers. If you missed several questions, you should probably reread the entire chapter carefully.

1. (a) support, blood formation, mineral reserve
   (b) support, protection, blood formation, mineral reserve
   (c) support, mineral reserve
   (frame 2)
2. They are trapped in the hardening of osteoid produced by osteoblasts. (frame 3)
3. (a) long; (b) short; (c) flat; (d) long; (e) irregular (frame 2)
4. (a) appendicular; (b) axial; (c) axial; (d) appendicular (frame 4)
5. (a) parietal; (b) temporal; (c) sphenoid; (d) maxilla (frames 6 and 8)
6. Cranium: frontal, occipital, ethmoid, sphenoid
   Face: vomer, mandible
   (frame 9)
7. (a) occipital (frame 7)
   (b) os coxae (frame 19)
   (c) scapula (frame 14)
   (d) tibia (frame 22)
   (e) ulna (frame 16)
   (f) sternum (frame 13)
   (g) ribs and sternum (frame 13)
8. os coxa (1), femur (1), patella (1), tibia (1), fibula (1), tarsals (7), metatarsals (5), phalanges (14) (frames 21 and 24)

CHAPTER FOUR
The Muscular System

The bony skeleton you studied in the last chapter provides a framework for the human body. Without muscles, however, it is not much use. The skeletal muscles of the body attach to the bones, using them as levers, to enable us to move. Other types of muscle support internal organs and pump blood through the body. In this chapter you will discover the different types of muscle, and will learn to recognize from the names of muscles many of their characteristics. Specifically, when you complete your study of this chapter, you will be able to:

- list and describe the three types of muscle cells;
- name and locate the three levels of connective tissue coverings in skeletal muscle;
- differentiate between the origin and insertion of a muscle;
- locate or describe a muscle, given a descriptive anatomical name;
- define terms describing muscle actions;
- locate and give the functions of certain muscles.

MUSCULAR TISSUE

1. Muscular tissue is specialized for contraction. This specialization, called contractility, means that muscle cells can get shorter with appropriate stimulation. When the stimulation is over, the muscle cells relax and assume their extended size.

   Three general types of muscle are present in the body. Skeletal muscle is attached to the skeleton of the body. These muscles are under voluntary control. You can move your arms at will, for example, or wrinkle your forehead. Cardiac muscle, which makes up the heart, is similar to skeletal muscle when examined under a microscope. It is not under voluntary control, however. Smooth muscle is likewise not under voluntary control, and is smooth in appearance, in contrast to both skeletal and cardiac muscle. Smooth muscle is found in the walls of most internal organs and tubes, such as blood vessels.

(a) Which two types of muscle are not under voluntary control? ________

____

(b) Which two types of muscle are not smooth in appearance under the microscope? ____________________

(c) Consider a muscle cell at rest. When it is stimulated will it get longer or shorter? ____________________

- - - - - - - - - - - - - - - - - -

(a) cardiac, smooth
(b) skeletal, cardiac
(c) shorter (contractility means it contracts or gets shorter)

2. Skeletal and cardiac muscle are striated or striped in appearance under the microscope, while smooth muscle is not. The striations represent regular recurrences of contractile fibers called myofibrils, within the muscle cell, which is also called a muscle fiber. When the muscle contracts, the myofibrils slide in a regular way, maintaining the appearance of cross banding in any stage of contraction. In smooth muscle, the myofibrils are not so regularly arranged; thus no cross banding is evident.

Smooth muscle cells are spindle-shaped with a single, centrally located nucleus. Skeletal muscle cells are very long, each with many nuclei. These nuclei are located at the edge of the cell, at various points along its length. Cardiac muscle cells are columnar in shape, sometimes branched, and each cell has a single, central nucleus. Cardiac muscle cells are joined end-to-end at specialized connections called intercalated disks. Skeletal and smooth muscle cells are generally connected side-by-side with connective tissue wrappings.

(a) Fill in the chart below.

Muscle type	Cell shape	Nucleus no., location	Striations yes/no	Control vol./invol.
Skeletal	________	________	________	________
Cardiac	________	________	________	________
Smooth	________	________	________	________

(b) What structures within a smooth muscle cell cause it to shorten?

____

(c) Why do the two types of involuntary muscle cells differ in the presence of striations? ______________________________________

____

(d) Which type of muscle cell is joined end-to-end? ______________

What are these end-to-end connections called? ______________

- - - - - - - - - - - - - - - - - - -

(a)

skeletal:	very long	many, at edges	yes	voluntary
cardiac:	columnar	one, central	yes	involuntary
smooth:	spindle	one, central	no	involuntary

(b) myofibrils (contractile fibers in all muscle cells)
(c) in smooth muscle the myofibrils are not as regular as in cardiac muscle
(d) cardiac, intercalated disks

In later chapters we shall see more of the characteristics of smooth and cardiac muscle tissue. In the rest of this chapter, we will be concerned with the skeletal muscles of the body. The muscular system ordinarily refers to skeletal muscle and its relations to the bony framework of the body.

3. Skeletal muscle forms much of the flesh of the body, covering the skeleton. Each "muscle," or bundle of muscle cells, functions as an organ. The muscle cells, as noted earlier, are very long. Each one contains very many myofibrils for contraction. Mitochondria are numerous in muscle cells as they produce the energy needed for contraction. A network similar to smooth-surfaced endoplasmic reticulum is also present to transmit the stimulation for contraction to all parts of the cell; this is the sarcoplasmic reticulum (sarco means muscle).

   Each long, thin muscle cell is called a muscle fiber and is covered with a cell membrane called the sarcolemma, as well as by a thin connective tissue sheath called endomysium. A bundle of several muscle fibers is wrapped in a fibrous connective tissue sheath called perimysium. Several of these bundles grouped together form what we call a "muscle"; it is wrapped by a third connective tissue covering called epimysium.

(a) On this diagram of part of a skeletal muscle, identify:

A ____________________

B ____________________

C ____________________

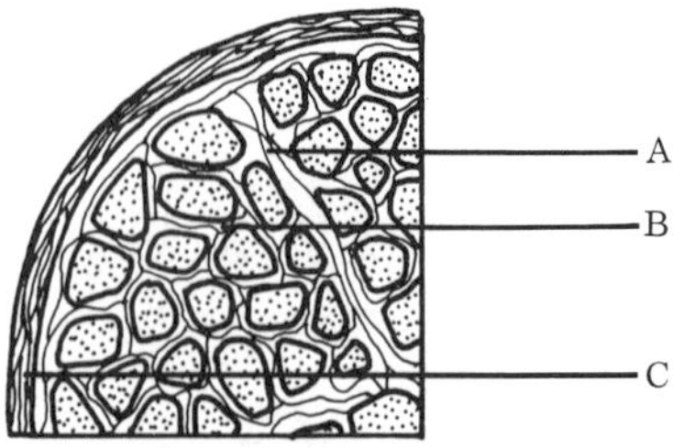

A
B
C

(b) On this diagram of a single muscle fiber, identify:

A ____________________

B ____________________

C ____________________

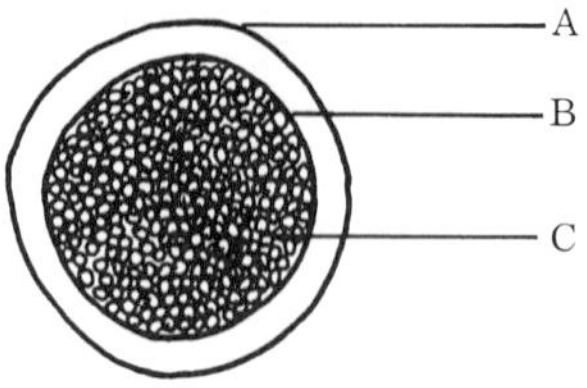

A
B
C

(c) What would be the function of a structure in muscle cells called sarcoplasmic reticulum? ____________________

(d) What organelle is quite numerous in muscle cells (besides microfilaments)? ____________________

- - - - - - - - - - - - - - - - - - - -

(a) A—perimysium; B—endomysium; C—epimysium
(b) A—endomysium; B—sarcolemma (or cell membrane); C—myofibril (or contractile filament)
(c) to transmit stimulation
(d) mitochondria (for energy)

SKELETAL MUSCLES

4. Skeletal muscles are attached to bone by tendons, which are bundles of heavy connective tissue fibers. The tendons attach at one end to the muscle and at the other to bone. When talking about attachments of muscles to bone, the tendons are generally ignored, but they are present in all cases.

   Skeletal muscles are generally attached to at least two bones. The less movable attachment of a muscle is called the origin of the muscle; thus, the muscle is said to originate on the less movable bone. The more movable bone is called the insertion; the muscle is said to insert in the more movable bone. A single muscle may have several points of origin. These "heads" combine in the belly, or fleshy part, of the muscle. Muscles are often named from the points where they attach to the bone, or by the number of heads of origin.

(a) The temporalis muscle is used in chewing. It is attached at one end to the mandible and at the other to the temporal bone. Which is the origin? ____________________ Which is the insertion? __________ ____________________

(b) The quadriceps femoris muscle is named by the number of heads and its general location. How many heads has it? ______________
Are these heads of origin or insertion? ____________________
In what area of the body is this muscle found? ______________ ____________________

(c) The sternocleidomastoid muscle in the anterior lateral neck has two origins and inserts into the mastoid process of the temporal bone. Contraction of this muscle bows the head. The origins are in the sternum and clavicle. What bone(s) is(are) more movable in this case? ____
What bone(s) is(are) less movable? ____________________

(d) Why isn't it precise to say the sternocleidomastoid muscle is attached to the sternum, clavicle, and temporal bones? __________ ____

- - - - - - - - - - - - - - - - - -

(a) temporal bone is origin; mandible is insertion
(b) four heads; of origin (on ilium and femur); found in area of femur (the insertion is on the tibia)
(c) temporal is more movable; sternum and clavicle are less movable (In this case, the entire head would bend, not just the temporal bone.)
(d) tendons are actually attached to the bones

Movements

5. Muscles are also often named for their general location. The Latin names of bones may be part of a muscle's name, as in the quadriceps femoris. The general term referring to the upper arm is brachialis, or brachii. Thus, the biceps brachii is a muscle of the upper arm, and has two heads of origin.

The effect of a muscle may also be part of a name. Two major muscle actions are flexion and extension. A flexor muscle decreases the angle between two bones, while an extensor muscle increases it. When you bend your elbow, you flex your arm. When you straighten it again, you use extensor muscles.

(a) A muscle in the forearm is called flexor carpi ulnaris. What bones would you expect to be involved? ________________

Which is the more movable portion? ________________

What action will this muscle cause? ________________

(b) The muscle name extensor indicis describes the effect of the muscle on a particular finger. What effect on what finger? ________________

(c) The extensor digitorum brevis and flexor digitorum brevis have origins on tarsals and insertions in phalanges. What is the differences in their effects? ________________

(d) Two sets of muscles are called the external intercostals and the internal intercostals. In what area of the body would you expect them to be located? ________________

Where would they be located relative to each other? ________________

- - - - - - - - - - - - - - - - - -

(a) carpals (wrist bones) and ulna; wrist; flex or bend the wrist
(b) extend or straighten the index finger
(c) the extensor straightens the toes, while the flexor bends them
(d) in the chest area between the ribs; the external intercostals would be closer to the skin and the internals would underlie the externals

6. Our bodies are capable of more movements than simple flexion and ex-extension. A part can be moved away from the midline (abduction) or brought back to it (adduction). A part may be rotated either toward the midline (medial rotation) or away from it (lateral rotation).

We noted some examples in the previous frame of flexor/extensor pairs of muscles. These new movements likewise often require a pair of muscles. Since muscles act by contraction, a second muscle is required to replace a moved part in its original position.

(a) What type of muscle action would be required to raise the arm straight out to the side? ________________

What type would lower it slowly? ________________

(b) The adductor brevis has its origin on the pubis and inserts into the femur. What effect would contraction of this muscle have? ________________

(c) What would be the effect of the levator ani muscle contraction on the anus? ______________________________

(d) What would be the effect of contraction of the depressor labii inferioris? (labii refers to lip) ______________________________

____

(e) The region of the buttocks is sometimes called the gluteal region for several large gluteal muscles located there. The gluteus maximus is used when you straighten your thigh relative to the hip and when you turn your leg so the toes and knee point away from the midline. What two muscle actions does gluteus maximus perform?

____

- - - - - - - - - - - - - - - - - - -

(a) abduction; adduction
(b) to adduct thigh (that is, to move it out to the side)
(c) to move it superiorly (that is, to raise it)
(d) to move the lower (inferior) lip downward (depressor)
(e) extending thigh and rotating it laterally

Muscles can be named for their general shape, size, or position relative to other muscles. These names will become apparent in the rest of this chapter, as you study a few important muscles as they relate to the bony skeleton you studied in Chapter 3 as well as to material to be covered in later chapters.

7. A few important muscles of the head and neck are indicated in the figure on the following page. The muscle surrounding each eye is orbicularis oculi, while that surrounding the mouth is orbicularis oris. The masseter muscle is vital for chewing. The platysma muscle forms a thin sheet over much of the neck, shoulder, and mandible, pulling down the corners of the mouth. The sternocleidomastoid has origins on the sternum and clavicle, with insertion on the temporal bone. If the right and left sternocleidomastoid work together, the neck flexes. If only one sternocleidomastoid muscle contracts, that side of the head moves inferiorly while the face is turned toward the opposite side in a rotating motion.

   (a) What would be the effect of contraction of the following muscles?

      (1) platysma muscles of both sides ______________________________

      (2) orbicular oris ______________________________

      (3) left sternocleidomastoid ______________________________

      (4) masseter ______________________________

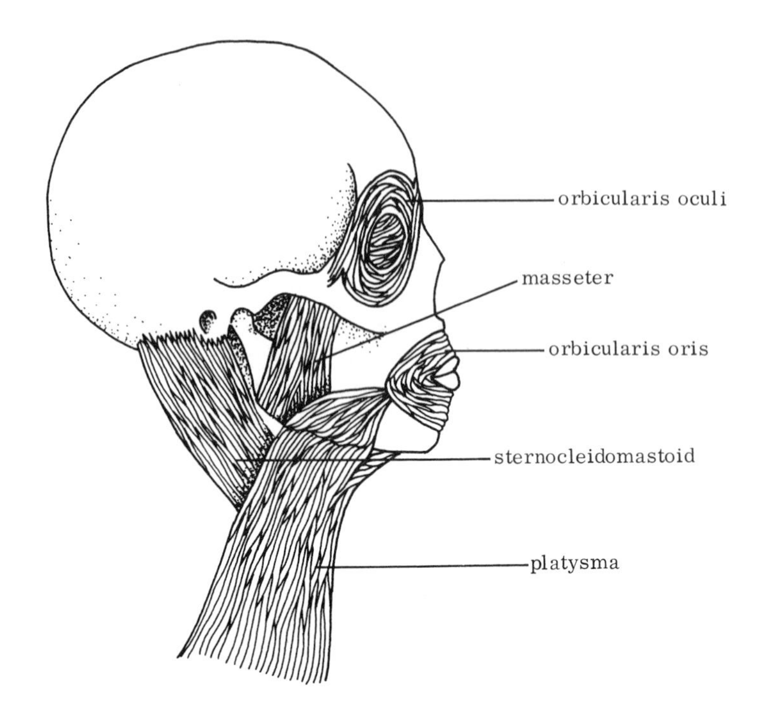

(b) Which muscle would function in closing the eyes? ______________

(c) Turn your head to the right and feel behind your left ear. Which muscle do you feel? ____________________ Follow the end down with your fingers. Where does it seem to end? ________________

____

- - - - - - - - - - - - - - - - - -

(a) (l) to pull down the corners of the mouth; (2) to pucker the lips; (3) to rotate face to the right; (4) to chew

(b) orbicular oculi

(c) sternocleidomastoid (left); you may be able to feel both points of insertion, one just to the left of the manubrium at the top edge of the breastbone, and the other a few inches down the clavicle

8. In the thoracic region the external and internal intercostal muscles are vital to breathing. Their contraction decreases the size of the thorax. The diaphragm also is critical in this regard. This large dome-shaped muscle separates the thoracic and abdominal cavities. When the diaphragm contracts, it pulls downward, increasing the size of the thoracic cavity and allowing more air to enter. Thus, the diaphragm contracts while the intercostals relax for the largest amount of air to enter the lungs.

The pectoralis major and minor assist in shoulder movements. Just as the biceps brachii flexes the forearm, triceps brachii extends it. Many smaller muscles function in the many separate movements of the wrist and fingers. The abdominal wall muscles act in conjunction with spinal muscles to extend, flex, adduct, abduct, and rotate the spine. Three layers of muscle in the abdominal wall are the external oblique (meaning fibers run diagonally), internal oblique, and the transverse abdominal (fibers run horizontally).

(a) Identify each of the following as being located in the thoracic, shoulder, or abdominal region.

(1) intercostals ____________________

(2) external oblique ____________________

(3) diaphragm ____________________

(4) pectoralis minor ____________________

(b) What are the effects of biceps brachii and triceps brachii? ________

____

(c) Which muscle becomes evident when you "make a muscle"? _______

____

(d) Suppose you are acting in a play and get "shot" early in the third act. You manage to breathe for twenty minutes without any evident chest movement. What muscle are you using to breathe? ______________

- - - - - - - - - - - - - - - - - - - -

(a) (1) thoracic; (2) abdominal; (3) between thoracic and abdominal; (4) shoulder
(b) biceps brachii flexes the forearm; triceps brachii extends it
(c) biceps brachii
(d) diaphragm

9. The muscles of the hip, thigh, and lower leg are many. We have already mentioned a few. The gluteus group makes up most of the flesh of the buttock, with gluteus maximus, gluteus medius, and gluteus minimus. These are used in such activities as bicycling and stair climbing. Biceps femoris and triceps femoris have functions similar to their counterparts in the upper arm. The sartorius muscle is usually the longest muscle in the average body; it extends from the anterior superior portion of the iliac bone to the medial head of the proximal portion of the tibia. The large muscle in the calf of the leg is the gastrocnemius; it extends from the distal femoral condyles to the calcaneus in the heel. The gastrocnemius aids in flexing the knee, as well as flexing the foot.

(a) An intermuscular injection ("shot") is usually given in the buttock. Which muscle group receives the medication? ____________

(b) The sartorius extends from the iliac bone to the proximal tibia. What actions would you expect from contraction of this muscle?

____

(c) Name the large muscle in the calf. ____________

(d) What is the effect of contraction of triceps femoris? ____________

____

- - - - - - - - - - - - - - - - - -

(a) gluteus group; (b) flexing of hip and knee; (c) gastrocnemius; (d) extend thigh

10. Give the location and function of each of the following muscles.

(a) platysma ____________

(b) gastrocnemius ____________

(c) internal intercostals ____________

(d) transverse abdominals ____________

(e) pectoralis major ____________

(f) masseter ____________

- - - - - - - - - - - - - - - - - -

(a) neck; pulls down corners of mouth
(b) calf; flexes knee
(c) between ribs; breathing
(d) abdominal wall; bends spine
(e) thoracic region; shoulder movement
(f) cheek; chewing

SELF-TEST

This Self-Test is designed to show how well you have mastered this chapter's objectives. Answer each question to the best of your ability. Correct answers and review instructions are given at the end of the test.

1. Muscle tissue is found in the walls of the heart.
   - (a) Is it striated or smooth? ____________________
   - (b) Is it voluntary or involuntary? ____________________
   - (c) What structures join two cells together? ____________________

2. Label the three levels of wrapping in this cross section of skeletal muscle.

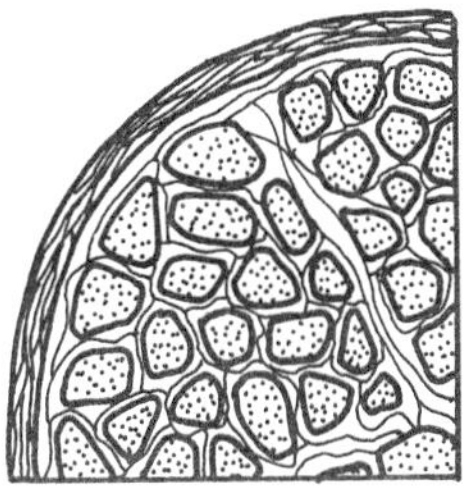

3. The extensor hallucis longus extends from the anterior fibula to the top of the great toe.
   - (a) What is its origin? ____________________
   - (b) What is its action? ____________________

4. The abductor pollicis brevis extends from a carpal to the first phalanx of the thumb.
   - (a) What is its insertion? ____________________
   - (b) What action does it cause? ________________

5. Name the movements described below.
   - (a) turning head to right ____________________
   - (b) lowering chin to chest ____________________
   - (c) raising chin from chest to normal position ____________________
   - (d) raising humerus to extend straight out ____________________
   - (e) moving humerus from position described in (d) to one with arms extending anteriorly ____________________

For each muscle named in the following questions, give the indicated information.

6. Extensor carpi radialis longus

   (a) origin ____________________

   (b) insertion ____________________

   (c) location ____________________

   (d) action ____________________

7. Biceps brachii

   (a) location ____________________

   (b) function ____________________

8. Sartorius

   (a) location ____________________

   (b) functions ____________________

9. Sternocleidomastoid

   (a) origin ____________________

   (b) insertion ____________________

   (c) action ____________________

10. Orbicularis oris

    (a) location ____________________

    (b) function ____________________

Answers

Compare your answers to the Self-Test questions with those answers given below. If all of your answers are correct, you are ready to go on to the next chapter. If you missed any, review the frame indicated in parentheses following the answers. If you missed several questions, you should probably reread the entire chapter carefully.

1. (a) striated; (b) involuntary; (c) intercalated discs (frame 2)
2. 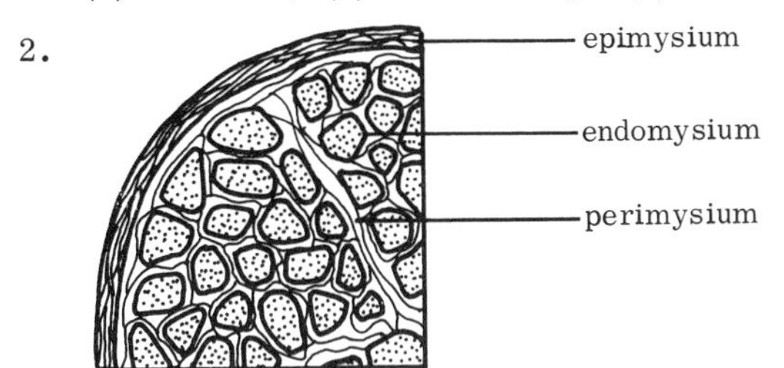

(frame 3)
3. (a) anterior fibula (less movable); (b) extend great toe (straighten it) (frames 4 and 5)
4. (a) proximal phalanx of thumb (frame 4)
(b) move thumb away from body (or palm) (frame 6)
5. (a) rotation (lateral); (b) depression; (c) elevation; (d) abduction; (e) medial rotation (frame 6)
6. (a) radius; (b) carpals; (c) wrist; (d) extend wrist (frame 5)
7. (a) upper arm; (b) flex elbow (frames 4 and 5)
8. (a) thigh; (b) flex hip and knee (frame 9)
9. (a) sternum and clavicle; (b) temporal bone (mastoid process); (d) flex or rotate head (frame 7)
10. (a) surround lips; (b) pucker lips (frame 7)

PART III

Communication Systems of the Body

In order for any living thing to survive, it needs an internal communication system. The parts of the human body must be aware of conditions in other parts. Suppose a bone is broken in the forearm. The muscles that normally move the forearm would continue to move the broken bone, resulting in further damage, if no communication took place within the body. In this case, however, the nervous system provides the communication—pain is produced when the arm is moved. This warns that something is wrong and the arm is generally left still.

Several body organs are strictly for communication. Your eyes and ears, for example, give information about the outside world to the brain, so your body can react to the situation. Many glands are also involved in communication. They influence the action of other body organs located far from the glands themselves. The production of substances by the glands is regulated by still other means of internal communication. Taken as a whole, the communication systems of the human body keep it functioning as a coordinated entity for an efficient existence.

CHAPTER FIVE

The Nervous System

Every living thing reacts to the world around it. In addition, all living things react to their internal environment. The human nervous system provides the internal communications network that enables and initiates the vital functioning of the body. After you complete your study of this chapter you will be able to:

- describe a typical neuron;
- explain how impulses are conducted along a nerve fiber;
- apply the terms afferent, efferent, sensory, and motor to nerves as appropriate;
- describe a simple reflex arc;
- specify the functions of various components of the nervous system;
- identify the components of the central nervous system;
- differentiate between the central and peripheral nervous systems;
- describe the major surface features of the brain.

NERVE TISSUE

1. The nervous system is very highly organized because of its function. It receives stimulations (stimuli) from both inside and outside the body. It transmits impulses from the source of stimulation to the central nervous system. There information is decoded, interpreted, stored, and used in sending impulses down effector nerve fibers to effect some change.

   The nervous system consists of a vast number of nerve fibers reaching all through the body, and a central control core which has been compared to the wires of a telephone regional headquarters.

   For study purposes, the nervous system is arbitrarily divided into the central nervous system (CNS), which includes the brain and spinal cord, and the peripheral nervous system (PNS), which includes all other neural components. Although we study it in two parts, the nervous system functions as a whole. All of the peripheral nerves arise or terminate in the central nervous system. Incoming and outgoing nerves can communicate with each other only through the central nervous system.

(a) Name the two divisions of the nervous system. ____________________

____

(b) Indicate to which of the divisions each of the following belong.

(1) brain ______________________

(2) nerve within leg ______________________

(3) spinal cord within thoracic vertebrae ______________________

- - - - - - - - - - - - - - - - - -

(a) central and peripheral

(b) (1) central; (2) peripheral; (3) central

2. Nerve tissue is basically made up of two types of cells: neurons and glial cells. Neurons are highly specialized to conduct impulses. All the neurons an individual will ever have are present when he or she is born. They grow in size but do not divide after that time. A typical neuron is shown below.

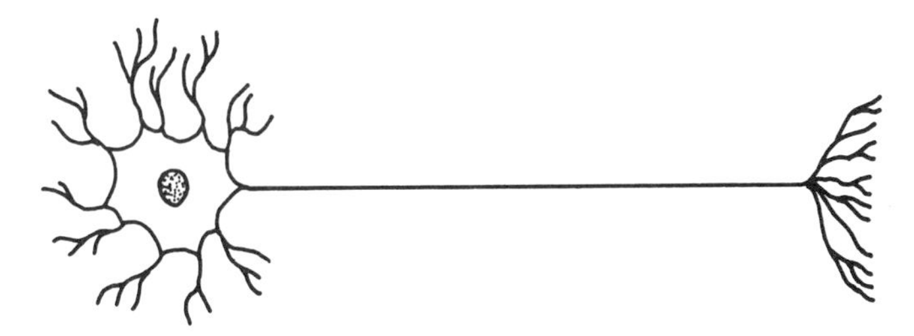

The cell body has an irregular shape. The long process is called an axon, and conducts impulses away from the cell body. The numerous short processes are called dendrites. These receive impulses and conduct them toward the cell body.

The cell body of a neuron contains many of the same organelles as other cells. Mitochondria, Golgi apparatus, nucleus with a nucleolus, and rough endoplasmic reticulum appear as discrete bodies under the light microscope. The RER was called "Nissl substance" before electron microscopy revealed it to be RER. Neurotubules (microtubules) are also found in neurons; these extend down the axon and carry substances that have some effect on the effector organ, that is, on whatever organ the axon terminates.

Glial cells support and nourish the neurons. They are of many different types, but we will look at only one in detail.

(a) Name the type of cell that is specialized to conduct impulses.

____

(b) What is the function of glial cells? _______________

(c) To what organelle in a neuron does "Nissl substance" refer?

____

(d) Which organelle extends down an axon? _______________

(e) In which direction does an axon conduct impulses? _______________

____

(f) What are the neuron processes other than an axon called? _______

____

- - - - - - - - - - - - - - - - - -

(a) neuron; (b) support neurons; (c) RER; (d) neurotubule; (e) away from cell body; (f) dendrites

3. (a) In the diagram in frame 2, which way would an impulse be transmitted along the axon—to the right or left? _______________

(b) What structure might transmit an impulse toward the cell body?

____

(c) How do products produced in the neuron cell body get to the distal end of the axon? _______________

- - - - - - - - - - - - - - - - - -

(a) to the right; (b) dendrite; (c) through neurotubules

4. When a nerve process is at rest, a negative charge exists inside it. The surrounding tissue has a positive charge. Impulses travel along neurons much as electricity travels along a copper wire. A stimulus starts the impulse, then the flow of ions (electrically charged atoms or molecules) into and out of the process causes a continuous change in the

charge of the membrane of the nerve process. This is called depolarization. The change in charge "flows" along as shown below.

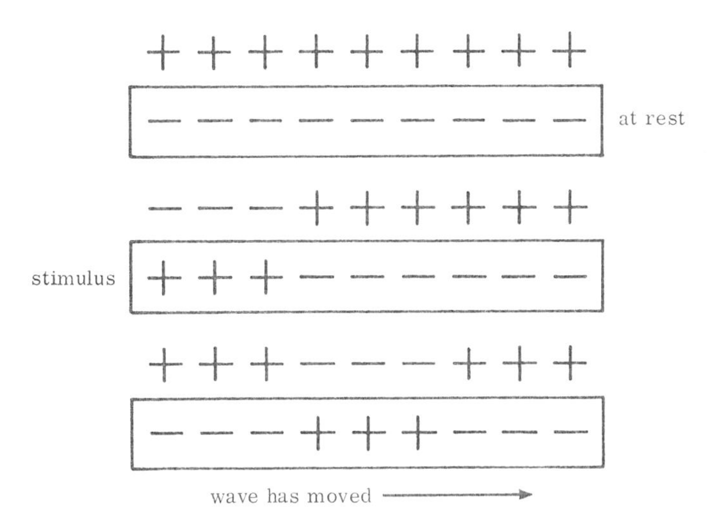

The impulse must be of a certain minimum strength to begin a depolarization wave. But once the minimum, or threshold stimulus, is reached, the wave does not lose strength or speed.

An impulse is passed from one neuron to another through a connection called a synapse. In the synapse an axon ending is near another axon, the dendrites of another neuron, or a cell body of another neuron. As an impulse reaches the axon terminal, chemical substances contained in vesicles are released into the gap between neuron endings. The chemicals, produced in the body of the neuron and transported to the end of the axon by means of the neurotubules, initiate a similar depolarization wave on the next neuron completing the impulse transfer through the synapse. While dendrites can receive one or many inputs, axons generally have a single transmission point.

(a) What is the threshold stimulus? ______________________________

____

(b) Across a synapse, is the impulse transmitted electrically or chemically? ______________________

(c) Where are the chemicals released into a synaptic space produced? ______________________ How do they get to the end of the axon?

____

- - - - - - - - - - - - - - - - - -

(a) minimum stimulus to initiate an impulse
(b) chemically
(c) in RER of cell body; via neurotubules

5. Neurons that activate muscles are referred to as motor neurons. The axon of such a motor neuron terminates on the surface of the muscle as a motor end plate, which includes both neural and muscular components.

As shown in the figure below, the axon terminates in a bulbous bouton, which fits into a synaptic trough on the surface of the muscle. As an impulse reaches the bouton, chemicals from vesicles are released into the synaptic trough. The chemical (acetylcholine) depolarizes the adjacent sarcolemma. The depolarization is rapidly spread to the entire muscle fiber through the sarcoplasmic reticulum we covered in the last chapter.

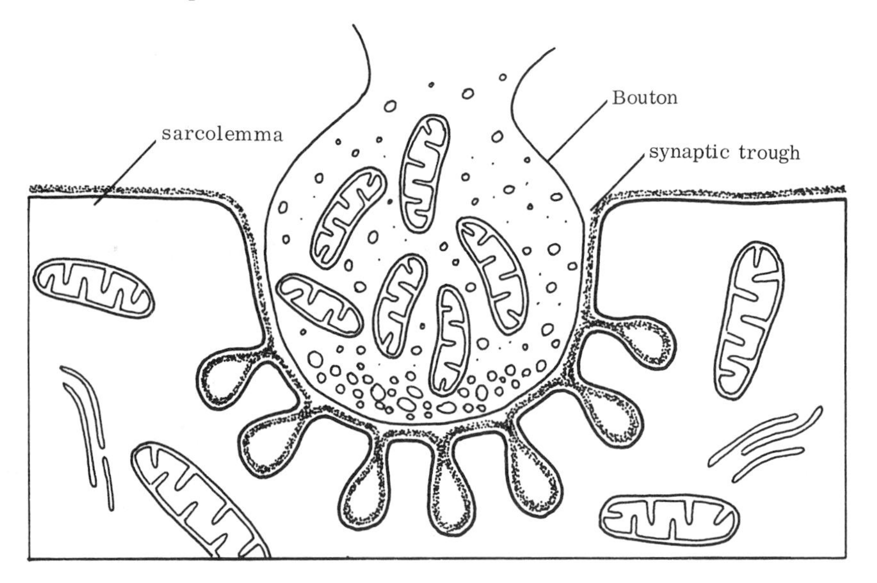

(a) What do you call the termination of a motor neuron? ______________________________

(b) In a myoneural junction, name the bouted end of the neural component. ______________________

(c) What is the muscular component of the junction called? ______________________________

- - - - - - - - - - - - - - - - - -

(a) motor end plate; (b) bouton; (c) synaptic trough

6. Processes of some types of neurons are each encased in a sheath made up of a glial cell known as the Schwann cell. Schwann cells form a thick myelin sheath around a segment of the neuron process during its early development, with the Schwann cell nucleus ending up in the outer layer of myelin. Many Schwann cells are required to ensheath a long fiber. The indentations on a myelinated fiber (see figure below) represent points between Schwann cells, and are called nodes of Ranvier.

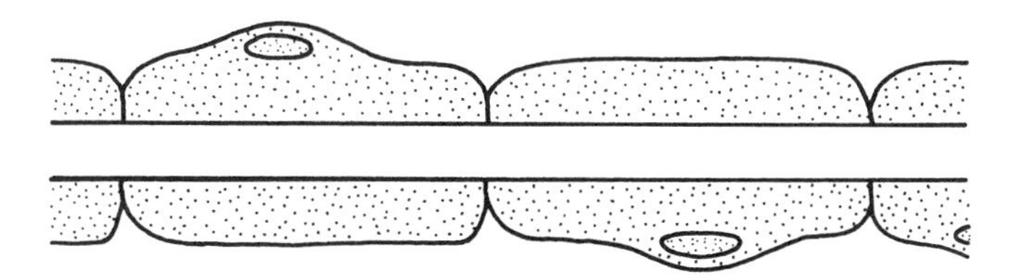

The myelin sheath speeds up conduction of impulses along a sheathed or myelinated nerve. The depolarization jumps from one node to the next in the form of conduction called saltatory conduction, rather than flowing along the neurilemma as described earlier. The saltatory conduction is faster than the flow. Impulses travel along unmyelinated fibers at a speed of 1 meter per second, but myelinated nerves conduct impulses as fast as 140 meters per second.

Recently, it has been verified that even fibers that do not have myelin sheaths are covered with a thin layer of Schwann cell cytoplasm. In these "unmyelinated" nerves, impulses are propagated in waves as described earlier.

(a) Both myelinated and unmyelinated nerve processes are covered by Schwann cells. Why is one thicker than the other? ______________

____

(b) How does impulse conduction differ in myelinated and unmyelinated nerves? ______________________________________

(c) Is the nerve represented below myelinated or unmyeliated? ________

____

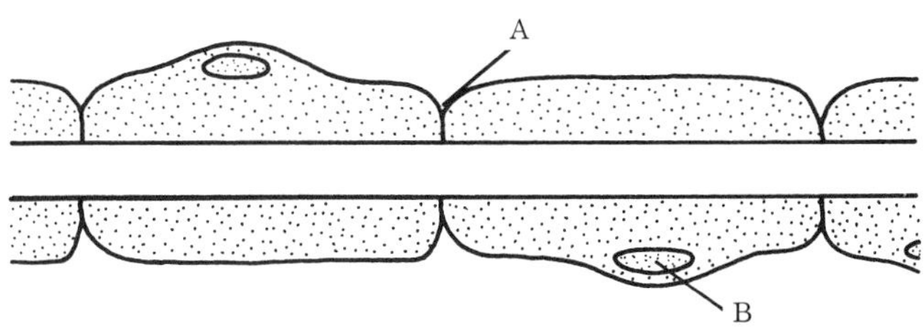

(d) What does A indicate? ______________________________

(e) What does B indicate? ______________________________

- - - - - - - - - - - - - - - - - - - -

(a) myelinated has several layers wound around, unmyelinated a single layer
(b) saltatory or jumping in myelinated, wave in unmyelinated
(c) myelinated
(d) node of Ranvier (junction of two Schwann cells)
(e) nucleus of a Schwann cell

7. In the central nervous system, the brain and spinal cord are also made up of neurons and glial cells. The cell bodies of neurons are grouped according to their functions. They are called nuclei (singular, nucleus) in the CNS or ganglia (singular, ganglion) when outside it. (Do not confuse a CNS nucleus, which is a group of neuron cell bodies, with the nucleus of a neuron or other cell.)

   Neurons totally within the CNS are called interneurons. Neurons conducting impulses toward the CNS are afferent, while those conducting impulses away from it are efferent neurons. (Afferent neurons approach the CNS; efferent neurons exit from it.) Afferent neurons are sensory; they receive a stimulus and transmit it to the CNS, which interprets it. Efferent neurons may be somatic (leading to an organ or gland) or motor (leading to a skeletal muscle). In either case, efferent neurons lead an impulse toward some effector organ. The effector organ may be a muscle or a gland, which translates the impulse into some action. Thus each neuron has a definite direction—impulses travel only on one-way streets.

(a) What is contained in a nucleus in the spinal cord? ________________

____

(b) What is contained in a ganglion? ________________________

(c) What is the difference between nuclei and ganglia? ________________

____

(d) What is the difference between afferent and efferent nerves? ______

____

(e) What is an interneuron? ________________________________

- - - - - - - - - - - - - - - - - - -

(a) neuron cell bodies
(b) neuron cell bodies
(c) nuclei are in CNS, ganglia are not
(d) afferent nerves conduct toward the CNS, efferent away from the CNS
(e) a neuron totally within the CNS

The reflex arc

8. The functional unit of the nervous system is the reflex arc, which includes parts of both the central and peripheral nervous systems. The reflex arc includes an afferent neuron, at least one interneuron, and an efferent neuron. The afferent neuron receives a stimulus and transmits it toward the CNS. At least one interneuron is needed to interpret the stimulus and synapse with an efferent neuron, which causes the reflex effect in some organ. For example, a receptor in the skin may be stimulated by a hot burner on a stove. The effect will be on skeletal muscles to move the stimulated area away from the heat.

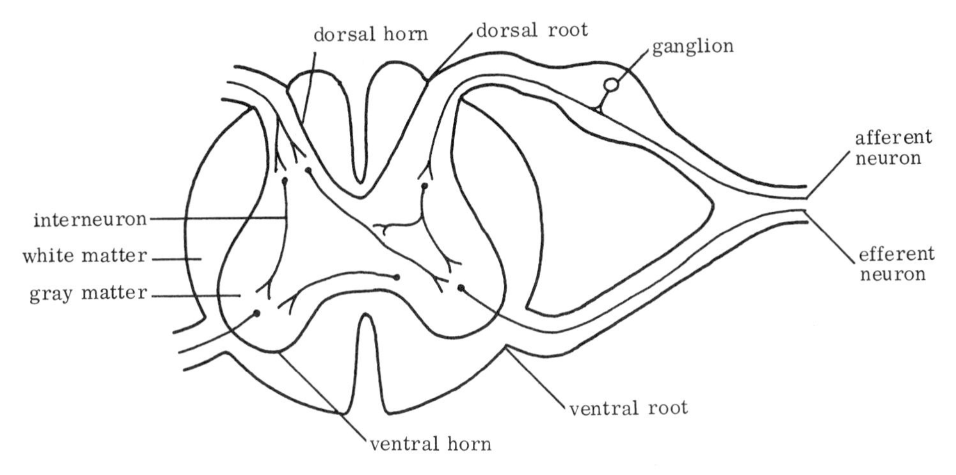

This diagram of a reflex arc shows also some of the structural features of the spinal cord. Dorsal (toward the back) and ventral (toward the front) horns can be seen in the gray matter of both sides. Roots are terms given to the junction of spinal nerves to the spinal cord. Each spinal nerve at each intervertebral space has these same features.

(a) What would be contained in the dorsal root ganglion? ____________

____

(b) What type of neuron passes through the ventral root? ____________

____________________ In the example of a reflex arc from burn to motion away from heat, what would be at the termination of this neuron? ____________________________________

(c) A neuron that runs from the dorsal to ventral horn is what type of neuron? ____________________

(d) What is the equivalent of a ganglion but within the spinal cord?

____

- - - - - - - - - - - - - - - - - -

(a) cell bodies of neurons
(b) efferent; motor end plate of skeletal muscle
(c) interneuron
(d) nucleus

9. In the cross section of a spinal cord at the right, the white matter is composed mainly of myelinated fibers. The gray matter consists of unmyelinated fibers and cell bodies.

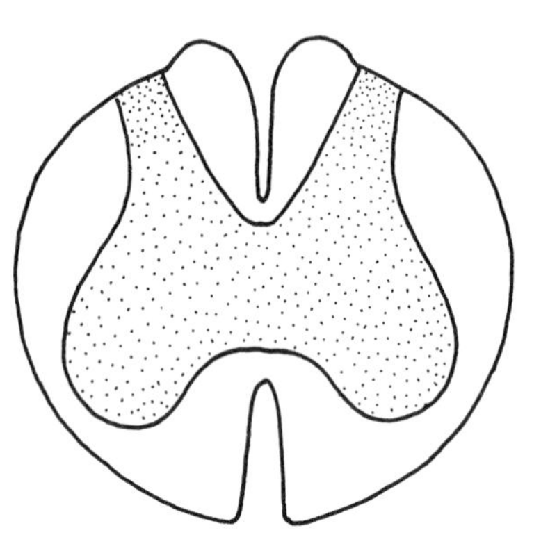

The gray matter has an H-shape with two dorsal and two ventral horns. Fibers are joined to the spinal cord at all horns. As you may have noticed in frame 8, the spinal nerves from dorsal and ventral horns on the same side join to form one nerve. This occurs on the right and left to form 31 pairs of spinal nerves.

Each spinal nerve, which is part of the peripheral nervous system, contains both sensory and effector nerves. Sensory neurons enter at the dorsal root with their bodies grouped into the dorsal root ganglia. Effector neurons located in the ventral horn have effector fibers that exit at the ventral root.

(a) Are ganglia associated with the dorsal or ventral root? ____________

(b) Do afferent fibers meet in the dorsal or ventral horn? ____________

(c) Would you expect to find myelinated or unmyelinated fibers in the horns of the spinal cord? ________________________

(d) Where would nuclei in the spinal cord be located? ________________

- - - - - - - - - - - - - - - - - - -

(a) dorsal; (b) ventral; (c) unmyelinated; (d) gray matter

10. A reflex arc involves components in the PNS, as well as structures in the gray matter of the spinal cord. It begins with stimulation of an ending in the peripheral nervous system, and ends with some action taken by another system, such as the muscular system. In the space below, list the required neural elements involved in a reflex arc. Be as specific as possible. Indicate also whether each element is part of the CNS or PNS. ____

____

- - - - - - - - - - - - - - - - - - -

(1) receptor (PNS); (2) afferent sensory neuron (PNS); (3) interneuron (CNS); (4) efferent motor neuron (PNS)

All of the peripheral nerves do not arise from the 31 pairs of spinal nerves. Some arise from the twelve sets of cranial nerves that provide still more fibers for the peripheral nervous system. Most of the cranial nerves supply parts of the head and neck. The exception is the tenth nerve pair, the vagus. This nerve supplies speech, swallowing, coughing, and vomiting as well as transmitting impulses from various thoracic and abdominal viscera, including the heart muscle. Several of the cranial nerves are specific to special senses such as seeing, hearing, smelling, or tasting. These will be treated in more detail in the next chapter. The cranial nerves arise from cell bodies located in the brain and brain stem. We will consider now these cranial structures.

THE BRAIN

11. The brain is defined as that part of the nervous system encased within the skull above the level of the foramen magnum of the occipital bone.

    The lowermost portion of the brain is continuous with the spinal cord and is called the medulla oblongata. This oblong structure controls several vital processes, such as heartbeat and respiration. Cranial nerves IX, X, XI, and XII (Roman numerals are used to name the cranial nerves) arise from the medulla oblongata. Four more cranial nerves arise from the pons, an ovoid bridge of tissue just superior to the medulla oblongata.

    The cerebellum is found posterior to the medulla and the pons. It is a large, spheroid organ with two lobes, and is concerned with regulating and coordinating muscular activity, although it does not initiate such activity.

    Superior and anterior to the cerebellum is the midbrain from which cranial nerves III and IV arise. Above this, the diencephalon, the source of the second cranial nerve, houses the vital hypothalamus, which regulates the activity of most abdominal organs. The hypothalamus is intimately related to the pituitary gland, which we shall investigate in Chapter 7. All of these structures except the cerebellum make up the brain stem.

    (a) The drawing on the following page shows the cerebellum and brain stem. The diencephalon at the superior portion of the brain stem is represented by A. Label the other areas of the brain stem, along with the cerebellum, using the locational information given in the text.

    (b) What is the primary function of the cerebellum? ________________

    ____

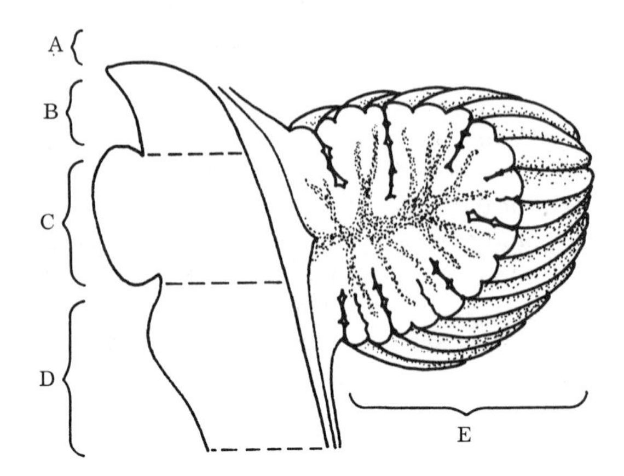

(c) What structure controls heartbeat and respiration? ______________

____

(d) In what structure is the hypothalamus? ______________________

(e) Which bone is present at the inferior border of the medulla oblongata? ______________________

- - - - - - - - - - - - - - - - - - -

(a) A—diencephalon; B—midbrain; C—pons; D—medulla oblongata; E—cerebellum
(b) to coordinate muscular activity
(c) medulla oblongata
(d) diencephalon
(e) occipital (foramen magnum)

12. The cerebrum, with two cerebral hemispheres, takes up most of the cranial activity. This structure is what most people would call the brain. The cerebral surface is convoluted, with deep folds called sulci (singular, sulcus), and fissures depending on their location. The longitudinal fissure separates the two hemispheres.

The diagram on the following page delimits the lobes of the cerebrum and the identifying fissures.

The central fissure (A) separates the frontal and parietal lobes. The area immediately anterior to the central fissure is concerned with sensation from most of the areas of the body, while the area just posterior (the post-central region) is the focus of corresponding motor activity. The lateral fissure (B) separates the temporal lobe from the frontal and parietal lobes. A less clearly defined parieto-occipital fissure separates the parietal and occipital lobes.

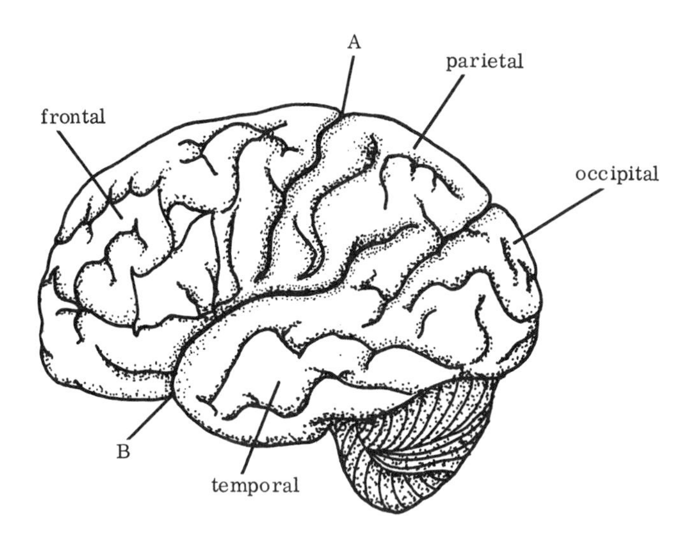

These lobes are named to correspond with the overlying cranial bones. The temporal lobe, as you might expect from its location, is the seat of hearing sensations. But the visual centers are located in the very back of the occipital lobe. The anterior portion of the frontal lobe is the association area—presumed to be involved in the thought processes of the "mind."

(a) On the figure below, label the areas listed here.

A—visual centers
B—hearing center
C—motor center
D—sensory center
E—association area

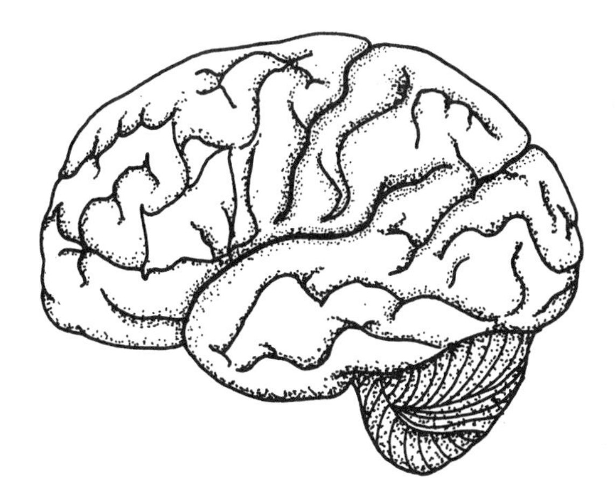

(b) Which lobes of the cerebrum are separated by the central fissure?

____

(c) Which fissure divides the two hemispheres of the cerebrum?

____

- - - - - - - - - - - - - - - - - - -

(a) See the figure as labeled.

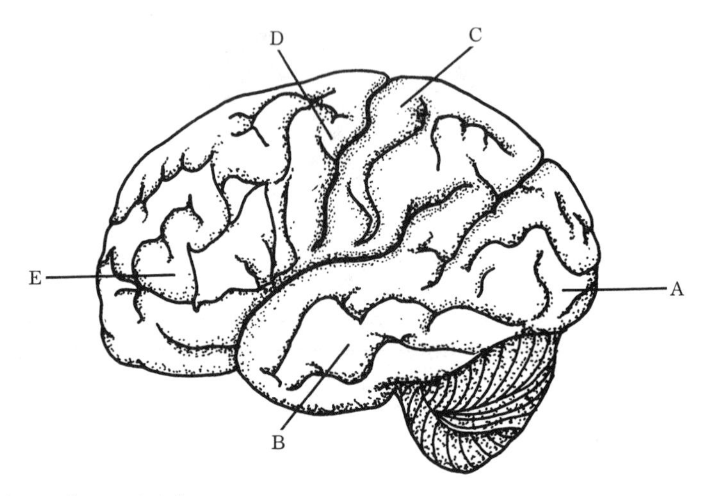

(b) frontal; parietal
(c) longitudinal

13. Give a function of each structure listed below.

(a) medulla oblongata ______________________

(b) cerebellum ______________________

(c) occipital lobe ______________________

(d) frontal lobe ______________________

(e) synapse ______________________

(f) neurotubules ______________________

- - - - - - - - - - - - - - - - - - -

(a) controls heartbeat and breathing
(b) controls muscular coordination
(c) controls vision
(d) thinking part of brain
(e) relays or transmits impulses
(f) transports products from cell body to axon end

14. The delicate structure of the brain and spinal cord are covered with layers of membranes called meninges, within the protective bone. The outer membrane is called the dura mater (hard mother); it is composed of fibrous connective tissue. The middle layer is called the arachnoid membrane for its spider-web appearance. The inner layer, the pia mater (gentle mother), is made up of delicate connective tissue fibers and cells with many small blood vessels. The pia mater follows the convolutions of the brain very closely.

    Any space between the layers, as well as spaces called ventricles within the brain, is filled with cerebrospinal fluid as a further protective device. The spinal cord terminates in the upper lumbar region, and the space below this is also filled with cerebrospinal fluid. Below this level, fluid can be withdrawn with a syringe for study with little likelihood of damaging the spinal cord.

    (a) What are the three meninges of the CNS? ____________________

    ____

    (b) Name three areas where cerebrospinal fluid may be found. _______

    ____

    (c) Which of the meninges is most vascular? ____________________

- - - - - - - - - - - - - - - - - - -

(a) pia mater (inner); arachnoid (middle); dura mater (outer)
(b) ventricles, below end of spinal cord, and spaces between meninges
(c) pia mater

AUTONOMIC NERVOUS SYSTEM

15. The autonomic nervous system (ANS) is a component of the peripheral nervous system. It consists of the motor nerves to smooth muscles of visceral organs. In fact, it supplies all effector organs except skeletal muscle.

    By means of the ANS, the body maintains its internal environment. Some of the activities made possible by the ANS are constriction and dilation of blood vessels, erection of the penis, "goose bumps," and secretion by glands. While afferent neurons do service the organs, they are not part of the autonomic system. In the ANS, nerve fibers from two opposed systems supply each organ. One of these fibers—the sympathetic fiber—can prepare the organ for a stressful or crisis situation. The other fiber—the parasympathetic—returns or maintains the organ in a normally effective condition. For example, in a crisis situation, sympathetic nerves bring about a rapid heartbeat. When the crisis has passed, parasympathetic nerves return the heart to its normal rate.

(a) Is the autonomic nervous system a division of the CNS or PNS?

____

(b) Does the autonomic nervous system consist of afferent or efferent fibers? ______________________

(c) How would autonomic fibers affect the rate of breathing? __________

____

(d) Sensory fibers from the stomach would be included in which nervous system division? ______________________________

- - - - - - - - - - - - - - - - - -

(a) PNS
(b) efferent only
(c) sympathetic fibers would speed it; parasympathetic would return it to normal
(d) PNS (the ANS contains only efferent fibers)

SELF-TEST

This Self-Test is designed to show how well you have mastered this chapter's objectives. Answer each question to the best of your ability. Correct answers and review instructions are given at the end of the test.

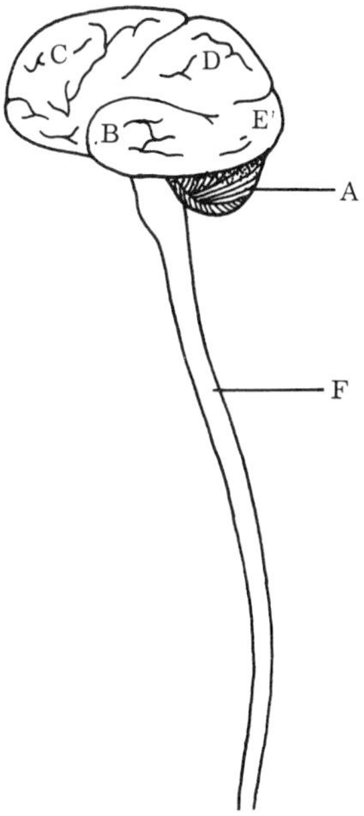

1. The diagram at the right represents which division of the nervous system? ____________ ______________________

2. Name the structure indicated by A. ________ ______________________

3. Which of the indicated structures would most likely be involved in a reflex arc? ________ ______________________

4. List in order the five components of a reflex arc, using the terms afferent, efferent, sensory, and motor to describe them. _____ ______________________ ______________________ ______________________

5. A part of the nervous system prepares the body to fight when a threat is sensed. What specific part of the peripheral nervous system does this? ______________________

6. What is the effect of Schwann cells on conduction of impulses along myelinated nerves? ______________________ ______________________

7. Answer the following questions about the diagram here.

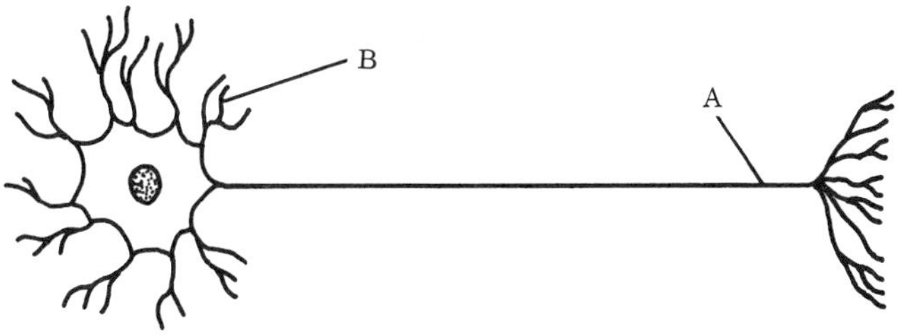

(a) What organelles might be found in A? ______________________

(b) What are the structures at B called? ______________________

(c) What makes up the Nissl bodies found in the cell body? __________

____

8. How are structures called ganglia and nuclei similar? ____________

____

9. Name a function carried out in the structures.

(a) cerebellum ____________________

(b) frontal lobe ____________________

(c) synapse ____________________

10. With what lobe of the cerebrum is each of the following associated?

(a) vision ____________________

(b) motor activity ____________________

(c) hearing ____________________

Answers

Compare your answers to the Self-Test questions with those answers given below. If all of your answers are correct, you are ready to go on to the next chapter. If you missed any, review the frames indicated in parentheses following the answers. If you missed several questions, you should probably reread the entire chapter carefully.

1. central nervous system (frame 1)
2. cerebellum (frame 11)
3. spinal cord (frame 8)
4. receptor; afferent sensory neuron; interneuron in spinal cord; efferent motor neuron; effector organ (frame 8)
5. sympathetic division of autonomic nervous system (frame 15)
6. cause saltatory or jumping conduction rather than wave motion, thus speeding it up (frame 6)
7. (a) neurotubules; (b) dendrites; (c) RER (frame 2)
8. both are collections of neuron cell bodies (nuclei are within the CNS and ganglia are outside it) (frame 7)
9. (a) muscular coordination (frame 11)
   (b) "thinking" (frame 12)
   (c) impulse transmission to another structure (frame 4)
10. (a) occipital lobe; (b) parietal lobe; (c) temporal lobe (frame 12)

CHAPTER SIX
The Special Senses

In the last chapter, we saw that the central nervous system integrates the various stimuli transmitted to it by the peripheral nervous system. In this chapter we will examine some of those stimuli and the organs and endings specialized to receive and code them for transmission to the CNS. Whenever you see, hear, smell, taste, or touch a portion of the outside world, these specialized endings are at work. When you complete this chapter you will be able to:

- list four elements critical to any sensation;
- identify two touch receptors in the skin;
- specify the role of a touch receptor in the reflex arc;
- label the component parts of the eye;
- distinguish between nearsightedness and farsightedness;
- trace the path of an impulse originating in the eye to the visual center in the brain;
- identify and locate the receptors of sight;
- specify the function of the component parts of the ear;
- explain how a sound wave gets from the outside world to a hearing receptor;
- locate the taste and smell receptors;
- differentiate between the mechanism of stimulation of taste and smell receptors.

In the preceding chapter we dealt in fairly general terms with the nervous system as a whole. Now we are going to examine in more detail the ultimate means by which we acquire information about our external environment. We will look at each of the five senses to discover just how the information is transmitted to the central nervous system for interpretation.

TOUCH

1. Receptors for the sense of touch are distributed throughout the skin. Touch is a fairly complex sensation, including different receptors for light touch, deep pressure, heat, cold, and pain. The endings for each of these sensations are different and are all attached to ends of dendrites. The sensation you feel depends, not on the type of stimulation, but on which nerve ending is stimulated. When an ending is stimulated, an impulse is generated that travels the afferent neuron toward the spinal cord. The result may be a simple reflex arc, or the impulse may be sent on to a higher center in the brain for interpretation. Following are the four critical elements of any sensation.

   Stimulus: some outside cause must be present.
   Receptor: some inside feature must respond to the stimulus.
   Pathway: the resultant impulse must go somewhere.
   Interpretation: the central nervous system must interpret the stimulus.

   Consider a reflex arc in which you pick up a hot spoon from a stove and instantly drop it.

   (a) What was the stimulus? ____________________

   (b) What type of receptor was stimulated? ____________________

   (c) What type of neuron provided the first pathway for the impulse?

   ____

   (d) Where was the impulse interpreted? ____________________

   - - - - - - - - - - - - - - - - - - -

   (a) hot spoon; (b) heat ending; (c) afferent; (d) in spinal cord

2. The nerve endings in the skin are generally of two types: encapsulated and free. The encapsulated endings have a covering or capsule of specialized connective tissue around the dendritic ending, while the free dendrite endings just terminate in the skin. Most of the free endings are pain receptors; there are more of these than all other nerve endings in the skin put together.

   The encapsulated endings vary for the different sensations. The largest encapsulated endings function in deep pressure, and are called Pacinian corpuscles. These corpuscles are 1 to 4 mm in length, and look like an onion in cross section. Meissner's corpuscles are smaller. and are the light touch receptors. They are ovoid structures, most numerous on the palms, the lips, and fingertips. Heat and cold sensations seem to be mediated by several types of endings, the best documented of which is the Krause end bulb. The Krause end bulb looks like a loose ball of string in its capsule, and seems to respond to cold stimuli. Overstimulation of any of the encapsulated endings can also produce a pain sensation.

(a) Name the ending that looks like an onion when cut. ______________ ______________ What stimulates this ending? ______________

(b) Name the ending that is stimulated by touch. ______________ Is it free or encapsulated? ______________

(c) What is the function of the Krause end bulb? ______________ What would result from too much stimulation of the Krause end bulb? ______________

(d) Describe a pain receptor. ______________

- - - - - - - - - - - - - - - - - -

(a) Pacinian corpuscle; pressure
(b) Meissner's corpuscle; encapsulated
(c) ending for cold; pain
(d) a free nerve ending; the tip of a dendrite

3. (a) Suppose you were trying to find your glasses in the dark. What nerve endings would you expect to be most helpful? ______________ ______________

(b) What nerve endings would aid you in determining if a bottle of wine were chilled properly? ______________

(c) Assume you are examining a section of skin from fingers under the microscope. At different powers, you see large, layered onion-like structures, and many free nerve endings. What would stimulate the encapsulated endings? ______________ What about the free ones? ______________

- - - - - - - - - - - - - - - - - -

(a) Meissner's corpuscles (hopefully, touch alone would do it)
(b) Krause end bulbs, the chief receptors for cold
(c) Deep pressure stimulates the encapsulated Pacinian corpuscle, while free nerve endings are associated with pain.

SIGHT

4. As the receptors in the skin respond to touch and temperature, the receptors in the eye respond to light and its patterns. The receptors for sight are located in the retina deep within the eyeball itself. The drawing below shows the component parts of the eye.

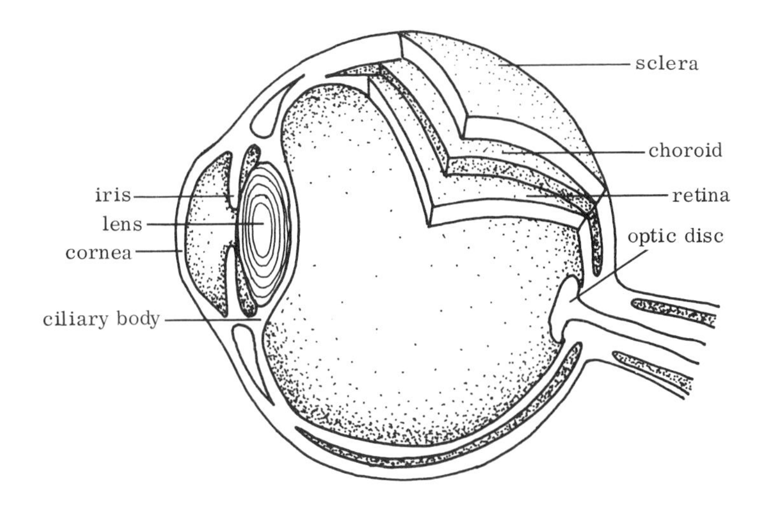

You can see that the wall of the eyeball has three layers. The sclera is the outer coat, comprising the white of the eye over most of the sphere and the clear bulging cornea in front. The middle layer is the choroid, which contains the blood vessels that supply the eye. In the anterior position, the choroid coat becomes the ciliary body, the muscle that adjusts the lens and iris for controlling light and focus. The inner layer is the retina; it contains receptors for sight called rods and cones. Rods and cones, which are stimulated by light, are endings from the optic, or second cranial nerve. The optic nerve enters the eye with blood vessels at the optic disc.

(a) In which layer of the eyeball are receptors for sight located? ______

______________________ What are these receptors called? ________

____

(b) The white of the eye is continuous with the cornea. What is it called? ______________________

(c) When light enters the eye it passes through two structures as well as fluid before it strikes the retina. What are these structures?

____

(d) Can you tell from examining the drawing what makes up the pupil of the eye? ______________________________

- - - - - - - - - - - - - - - - - -

(a) retina; rods and cones
(b) sclera
(c) cornea and lens
(d) the space around which is the iris; blank area in front of lens

5. As you read this page, the lens of each of your eyes focuses an image on its retina, just as, when you take a photograph, an image of what you see in the viewfinder is focused on the film. The nerve endings on the retina send impulses along the optic nerve to the visual center in the brain. Here the pattern is interpreted (as a photograph is developed) and we "see."

The pattern of impulses depends on the pattern of light and its colors that made up the stimulus and on which type of endings they stimulate. The cones in the retina are effective in bright light and for most color discrimination, while the rods function when light is dim. One spot in the eye, the fovea, contains only cones; this is the point of best vision. The optic disc, the entrance of the optic nerve and blood vessels into the eyeball, contains neither rods nor cones. It is often called the blind spot since no visual receptors are present.

(a) Do the receptors in the fovea of the eye function best in bright or dim light? ______________________

(b) What structures are found at the blind spot of the eye? ____________ ______________________

(c) Which receptors would be defective if a person were color-blind? ______________________

- - - - - - - - - - - - - - - - - -

(a) bright; (b) optic nerve and blood vessels—no rods or cones; (c) cones

6. The cornea and the lens work together to focus an image on the retina to stimulate the rods and cones. In the normal eye, this is accomplished as shown in the figure on the following page.

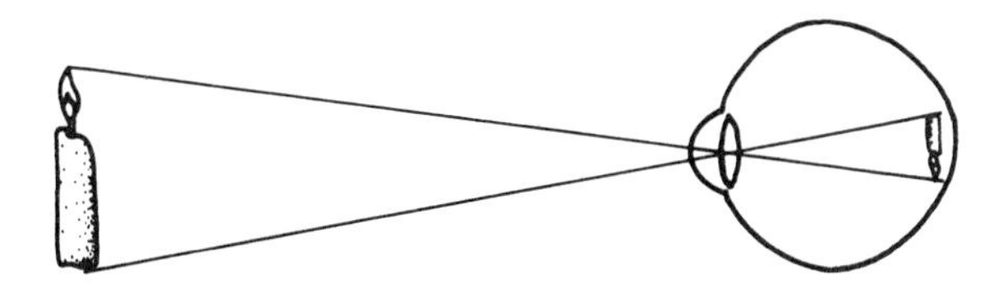

Many eyes, however, are not perfectly shaped. Whether the eyeball is shorter or longer, the resultant vision will be unclear as a result of imperfect stimulation of the receptors.

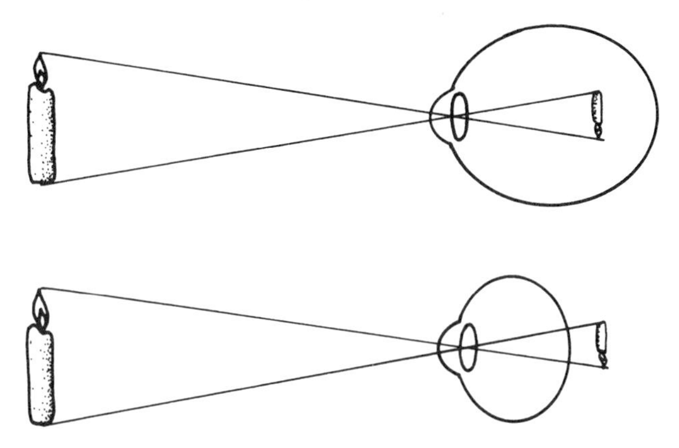

When the image is focused anterior to the retina, the condition is called myopia, and the individual is said to be nearsighted. He can see objects well only when they are close to him. When the focal point is behind the retina, the image on the retina itself is naturally out of focus or fuzzy. This condition is called hyperopia, and the individual is said to be farsighted. The farsighted person has better vision for distant objects. Both conditions can be corrected with lenses.

(a) A person who wears reading glasses, but does not need correction for driving, would probably have what condition? ______________

(b) Would someone who needs glasses for driving but never for close work be nearsighted or farsighted? ____________________

(c) In a normal eye with a perfect lens, where is the image focused?

____

- - - - - - - - - - - - - - - - - - -

(a) farsighted (he has hyperopia)
(b) myopia (he is nearsighted)
(c) on the retina (Other variations in eyeball shape cause other irregularities in vision. Many of these can also be corrected with appropriate lenses.)

7. Stimulation of the retina causes impulses to be transmitted to the visual center in the very posterior part of the cerebrum. The optic nerve exits each eye at the optic disc. The two optic nerves meet at the optic chiasma, then separate, as shown in the following figure. Most of the impulses from the right visual field (right sides) of each eye are eventually transmitted to the right occipital lobe, while most impulses from the left visual field of each eye are eventually transmitted to the visual center in the left occipital lobe. The visual impulses are then interpreted by other areas of the occipital lobe.

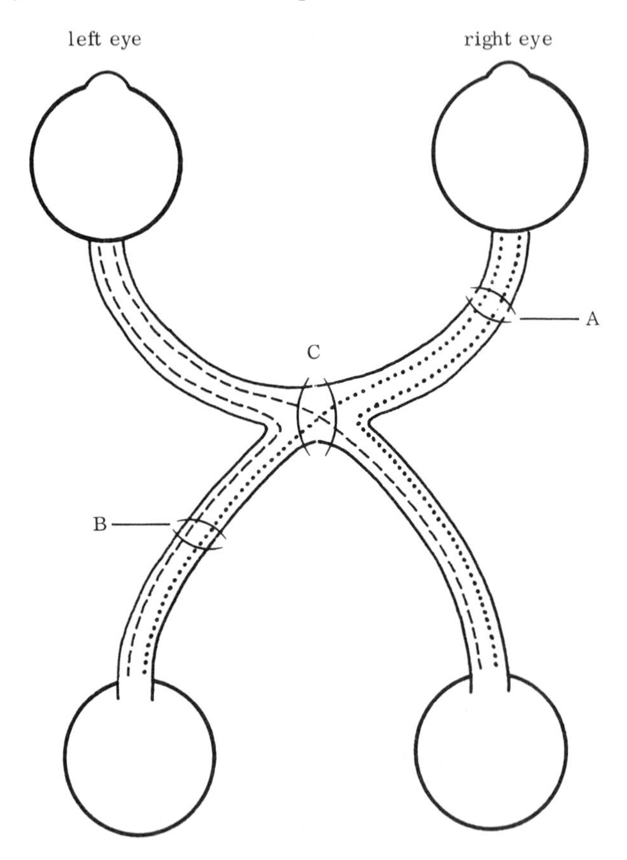

(a) Suppose the optic nerve were damaged or cut at point A in the figure. In which of the following areas would vision be most affected? Right visual field of right eye; Left visual field of right eye; Right visual field of left eye; Left visual field of left eye. ________

____

(b) If the optic nerve were cut at point B which areas would be most affected? ________________________

(c) Which areas would be affected by a cut through point C? ______

______

(d) What do you call the area at point C? ______

(e) In which lobe of the cerebrum are visual centers located? ______

______

- - - - - - - - - - - - - - - - - -

(a) both visual fields in the right eye; (b) left visual field in each eye; (c) right field in left eye and left field in right eye; (d) optic chiasma; (e) occipital

8. These questions summarize the sense of sight.

(a) Label the structures indicated in the figure below.

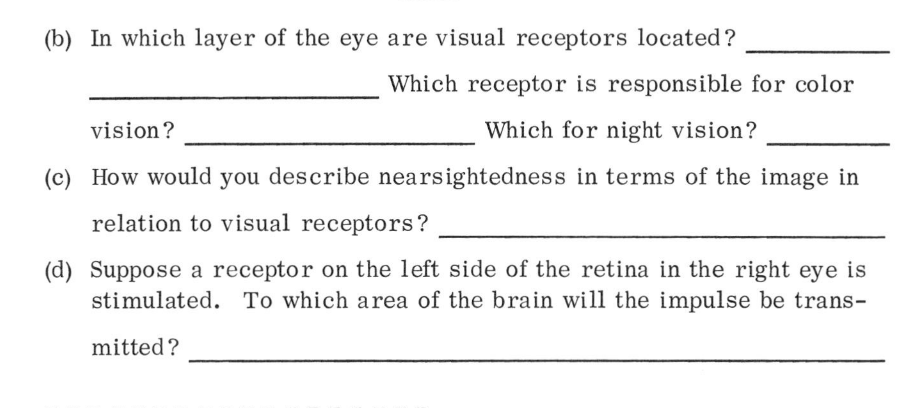

(b) In which layer of the eye are visual receptors located? ______

______ Which receptor is responsible for color vision? ______ Which for night vision? ______

(c) How would you describe nearsightedness in terms of the image in relation to visual receptors? ______

(d) Suppose a receptor on the left side of the retina in the right eye is stimulated. To which area of the brain will the impulse be transmitted? ______

- - - - - - - - - - - - - - - - - -

(a) A—iris; B—cornea; C—sclera; D—choroid; E—retina; F—optic disc
(b) retina; cones; rods
(c) image is focused in front of retina
(d) visual center in left occipital lobe

9. For the sense of sight, name each of the following.

Stimulus ____________________

Receptor ____________________

Pathway ____________________

CNS area ____________________

- - - - - - - - - - - - - - - - - - -

light; rods and cones; optic nerve; occipital lobe

HEARING

10. The sense organ for hearing is the ear and the auditory mechanism within. The auditory mechanism responds to the stimulus of sound. The waves in air created by sound are gathered by the outer ear, or auricle, and funneled into the ear canal (also called the external auditory meatus). At the end of the ear canal is the tympanic membrane or eardrum. The sound waves cause the eardrum to vibrate.

The middle ear is a small cavity just medial to the eardrum. A tube connects the middle ear to the throat; this auditory (Eustachian) tube equalizes pressure on both sides of the eardrum. Within the middle ear are three small bones, the ossicles, that transfer vibrations of sound waves from the eardrum to the membrane that masks the inner ear. The ossicles, the malleus, incus, and stapes, are connected in series with two small muscles to convert the air vibrations of the outer ear to fluid vibrations for the inner ear.

(a) Vibrations of what structure cause the ossicles to move? ____________________

(b) What sets off the first vibrations in the ear? ____________________

(c) What is the function of the ear canal? ____________________

(d) What is the function of the auditory (Eustachian) tube? ____________________

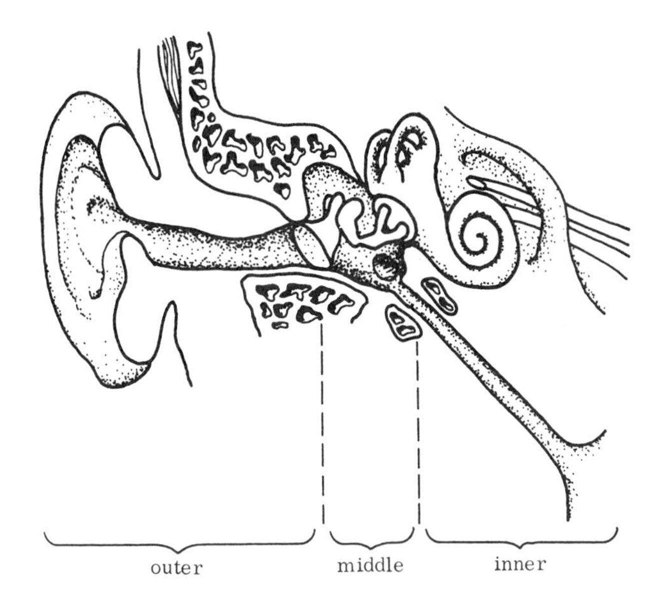

(e) In which division of the ear (outer or middle) would each of the following be found?

auricle ____________________

ear canal ____________________

ossicle ____________________

auditory tube ____________________

- - - - - - - - - - - - - - - - - -

(a) ear drum (tympanic membrane)
(b) sound waves
(c) funnel sound waves to the ear drum
(d) equalize pressure on both sides of the ear drum
(e) outer ear; outer ear; middle ear; middle ear

11. The inner ear contains the organ with hearing receptors, the shell-like cochlea. The end of the cochlea is covered by a membrane called the oval window, to which the stapes is attached. Movement of the stapes moves the membranous oval window, which creates waves in the fluid-filled cochlea. This in turn causes vibration of the basilar membrane which stretches along the cochlear lining. As the basilar membrane vibrates, hair cells of the organ of Corti are bent, causing impulses to be generated. These hair cells are sensitive to varying pitches. A membranous round window at the base (end farthest from the oval window) of the cochlea serves to damp the vibrations, after the receptors are

stimulated. As the oval window is pushed in by movement of the stapes, the round window bulges out at the base of the cochlea, to prevent any damage to the sensitive organ of Corti by the waves.

(a) What membranes separates the middle and inner ear? ____________

____

(b) To which of the two windows is the stapes attached? ____________

____

(c) The receptors for hearing are "hair cells" located in what organ?

____________________ On what membrane? ____________________

- - - - - - - - - - - - - - - - - -

(a) round window and oval window
(b) oval window
(c) organ of Corti (which is in the cochlea); basilar membrane

12. The inner ear contains two types of organs, the semicircular canals and maculae, that are involved not with hearing but with balance. The semicircular canals are fluid-filled structures that are so arranged to perceive any movement of the head. The maculae are also organs of position; small otoliths (ear stones) rest on hairs within the maculae to inform the brain of head position. The semicircular canals relay information about position in movement to the brain. The maculae are concerned with static or still position.

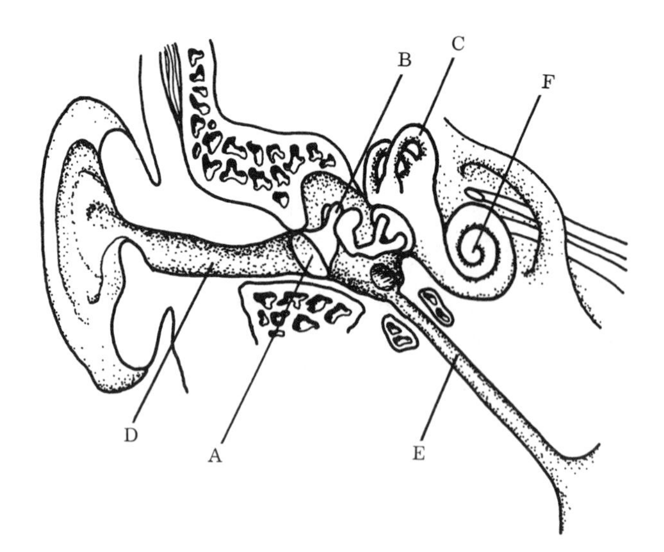

(a) Which structure in the figure contains the receptors for hearing?

____

(b) Which structures convert air waves to fluid waves? ____________

(c) Which structures respond to movement of the head? ____________

____

(d) Which structure equalizes pressure across the eardrum? ________

____

- - - - - - - - - - - - - - - - - -

(a) F; (b) B; (c) semicircular canals; (d) E

13. The impulses generated by receptors for sound travel over the eighth cranial nerve, the auditory or vestibulocochlear nerve. The fiber leads to an area in the brain stem. Some of the fibers cross over, as at the optic chiasma, so that impulses from both ears are represented in the auditory area of each temporal lobe.

(a) For the sense of hearing, name each of the following.

Stimulus ________________

Receptor ________________

Pathway ________________

CNS area ________________

(b) List (in order) the structures a sound wave passes from the outside air to the appropriate cranial nerve. ________________

____

- - - - - - - - - - - - - - - - - -

(a) sound waves; hair cells in organ of Corti; vestibulocochlear (acoustic nerve); temporal lobe
(b) auricle, ear canal, eardrum, ossicles, oval window, cochlea, receptors in organ of Corti, vestibulocochlear nerve (or auditory nerve)

SMELL AND TASTE

14. The senses of smell and taste are somewhat interrelated, as anyone who has engaged in much feasting with a stuffy cold can attest. They are similar in that both are chemical senses, whereas sight and hearing are

physical senses. Molecules of substances are necessary to activate smell and taste receptors. In both cases, the substance to be sensed must be dissolved in fluid; in smell, the substance must be volatile (airborne) as well.

The receptors for taste are located in taste buds in the oral cavity. They are most numerous in the tongue, but are also found in the soft palate and on the throat walls. These receptors resemble hairs in the taste buds, but are actually short dendrites, leading to three separate cranial nerves. Impulses from all three nerves pass into the parietal lobe for processing.

Olfactory (smell) receptors are located in the superior part of the nasal cavity. These receptors are neurons that extend from the lining of the cavity through the overlying bone into the olfactory bulb directly above. From here, the first cranial nerve, the olfactory, carries the impulses to olfactory areas in the frontal and temporal lobes.

Fill in the following chart for the chemical senses.

	Smell	Taste
Stimulus	________________	________________
Receptor (location)	________________	________________
Pathway	________________	________________
CNS area	________________	________________

- - - - - - - - - - - - - - - - - - -

Smell	Taste
volatile (airborne) dissolved molecules	dissolved molecules in fluid
neurons in top of nasal cavity	taste buds, dendrites resembling hair cells, mainly on tongue
direct to olfactory bulb, olfactory nerve	three cranial nerves
temporal and frontal lobes	parietal lobe

15. The taste buds in the tongue are sensitive to four basic tastes—salt, sweet, sour, and bitter. Recent studies show there may be more, such as specific receptors for water and electric (or metallic) tastes.

The receptors are not evenly distributed. The receptors for bitter taste are found on the back of the tongue as well as on the palate. Sweet taste is located on the tip of the tongue. Salty taste is found at the anterior sides while sour sensations are concentrated on the posterior sides of the tongue. There is, of course, some overlap but the pattern seems to be a universal one.

Odors have not been classified into a small number of groups, but seem to be based on molecular shape. The gaseous molecules are dissolved in mucous on the surface of the olfactory lining. They then stimulate the neurons in some way. A substance to be tasted must likewise be dissolved. Dry sugar, on the tongue for example, isn't tasted until saliva dissolves it. As in olfaction, the exact mechanism of stimulation is unknown.

(a) How does the source of stimulation differ for smell and taste sensations? ____________________

(b) Where are the receptors located that respond to volatile substances?

____

(c) Caffeine is classified as a bitter substance. On what part of the tongue would sensitivity to it be greatest? ____________________

- - - - - - - - - - - - - - - - - - -

(a) both must be dissolved, but for smell it must be volatile (in gaseous form; (b) superior nasal cavity; (c) back or posterior

SELF-TEST

This Self-Test is designed to show how well you have mastered this chapter's objectives. Answer each question to the best of your ability. Correct answers and review instructions are given at the end of the test.

1. What is the function of each of the following?

   (a) Pacinian corpuscle ______________________________

   (b) Meissner's corpuscle ______________________________

2. What are the three layers of the posterior portion of the eye? __________

   ____

3. What is the visual effect of myopia? ______________________________

4. Name the two types of visual receptors and indicate which is involved with color discrimination. ______________________________

   ____

5. An image interpreted in the left occipital lobe would have originated in one of two areas in the eyes. What are the areas? ________________

   ____

6. Indicate the order in which the following structures occur in the path of a sound wave.

   __________ organ of Corti

   __________ auricle

   __________ ossicles

   __________ eardrum

   __________ oval window

   __________ vestibulocochlear nerve

7. What is the function of each of the following?

   (a) ear canal ______________________________

   (b) ossicles ______________________________

   (c) semicircular canals ______________________________

8. Specify the location of each of the following.

   (a) rods ____________________

   (b) Krause end bulb ____________________

   (c) organ of Corti ____________________

   (d) olfactory receptor ____________________

   (e) taste buds ____________________

9. In general, how do the senses of sight and hearing differ from those of smell and taste? ____________________

10. Name four features that are necessary for any sensation. ____________________

____

____

Answers

Compare your answers to the Self-Test questions with those answers given below. If all of your answers are correct, you are ready to go on to the next chapter. If you missed any, review the frames indicated in parentheses following the answers. If you missed several questions, you should probably reread the entire chapter carefully.

1. (a) pressure; (b) touch (frame 2)
2. sclera, choroid, and retina (frame 4)
3. The image is focused in front of retina, thus the viewer can see only close objects. (frame 6)
4. rods and cones; cones are concerned with color (frame 5)
5. left visual field of left eye and the left visual field of right eye (frame 7)
6. 5; 1; 3; 2; 4; 6 (frame 10)
7. (a) to funnel sound waves to eardrum (frame 10)
   (b) to transfer air waves from eardrum to fluid waves at oval window (frame 10)
   (c) to sense movement of the head (frame 12)
8. (a) retina of the eye (frame 4)
   (b) skin (frame 2)
   (c) cochlea of inner ear (frame 11)
   (d) superior nasal cavity (frame 14)
   (e) tongue (frame 15)
9. Sight and hearing are physical senses while smell and taste are chemical senses. (frame 14)
10. stimulus, receptor, pathway, and CNS area (frame 1)

CHAPTER SEVEN

The Endocrine System

The glands of the endocrine system are located in various parts of the body. They do not form a continuous network as does the nervous system. Yet functionally the endocrine glands are more alike than different. For this reason, they are treated as a system. After you complete your study of this chapter, you will be able to:

- identify and locate seven different endocrine glands;
- differentiate functionally between exocrine and endocrine glands;
- specify the source and effect of each of nine pituitary hormones;
- explain the interaction between the hypothalamus and the pituitary gland;
- describe a thyroid follicle, and give its function;
- name the hormones produced by the pancreas, and compare their effects;
- specify the effects of adrenal medullary production;
- identify the stimulation and result of adrenal cortex production;
- relate testicular and ovarian hormone production to pituitary production;
- specify the form of control of production of the various endocrine hormones.

ENDOCRINE GLANDS

1. Many cells in the body secrete their manufactured products. The cells specialized for secretion are all derived from epithelial tissues. Many secretory cells remain in close association with epithelium. They have ducts lined with epithelium to conduct secretions to the appropriate location.

   Other epithelial derivatives do not remain connected to epithelium. Their secretions pass directly into the blood stream which carries them to the desired spot. The glands with ducts are called exocrine glands while the ductless ones are endocrine glands.

(a) The salivary glands produce saliva which is carried to the oral cavity through epithelium-lined structures. Are salivary glands endocrine or exocrine? ____________________

(b) The thyroid gland in the anterior neck secretes a product that is carried by the bloodstream. Is the thyroid gland endocrine or excrine? ____________________

(c) The pituitary gland is endocrine. Do its secretions pass through a duct or pass directly into the blood? ____________________

(d) The liver is an exocrine gland that produces substances used in the digestion of food. How are its secretions carried to the digestive organ? ____________________

- - - - - - - - - - - - - - - - - -

(a) exocrine; (b) endocrine; (c) they pass directly into the blood; (d) through a duct

2. The diagram on the following page shows the locations of endocrine glands in the human body.

Unlike other systems, the organs of the endocrine system are not physically adjacent or interconnected. Since their secretions all enter the blood, however, they can influence each other by this route. The endocrine glands function to integrate and coordinate body processes by means of chemical secretions called hormones.

The pituitary gland near the base of the brain is sometimes called the "master gland," as it influences many others. The thyroid gland and four parathyroid glands are located in the neck. Two adrenal glands are located atop the kidneys. Portions of the pancreas, in the abdominal cavity, are also considered endocrine in nature. The gonads, or sex glands, are found in or near the pelvic cavity. The ovaries in women or testes in men contribute to sexual characteristics of the person.

(a) Name the endocrine glands located in the abdominal region. ________

____

(b) Which endocrine gland is located nearest to the brain? ________

____

(c) How many endocrine glands are located in the neck region? ________

____

(d) The gonads secrete "sex hormones"; the pair present depends on the sex of the individual. Name the glands for each sex.

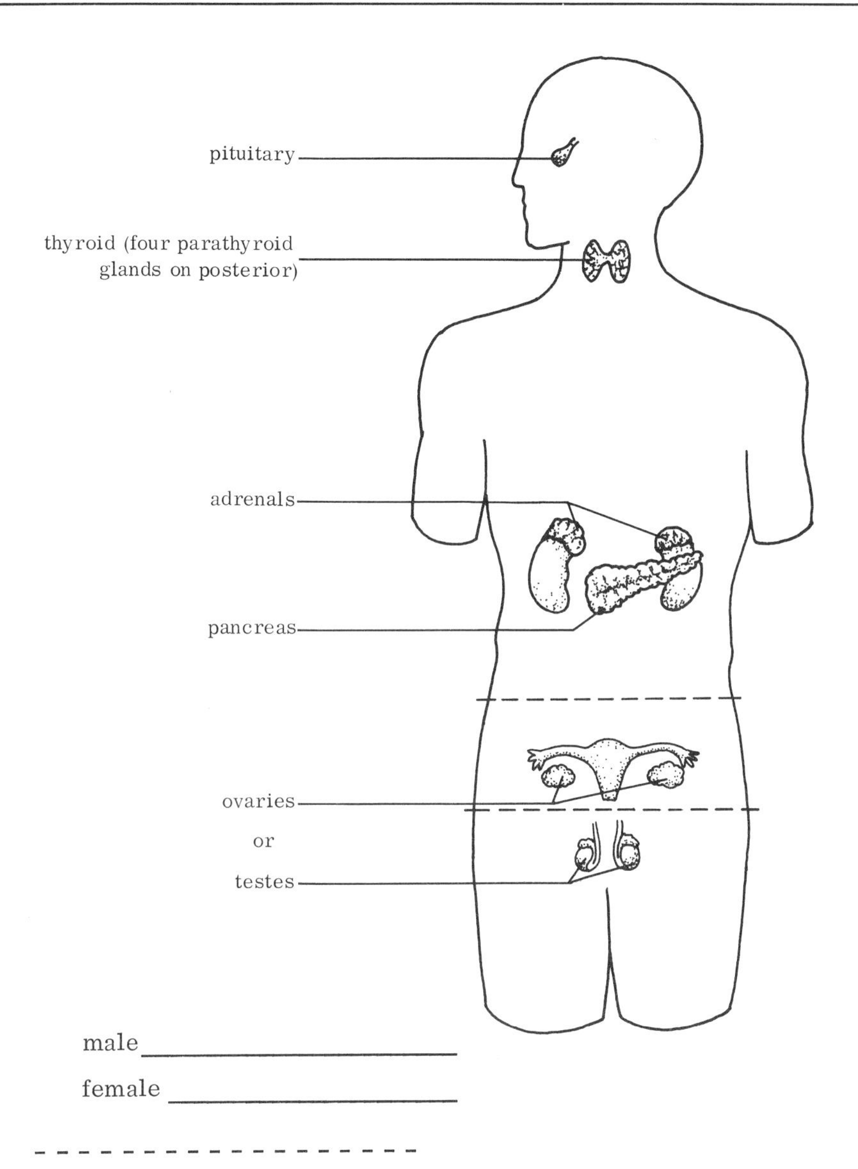

(a) right and left adrenal, pancreas
(b) pituitary
(c) five (thyroid and four parathyroids)
(d) testes, ovaries

3. The secretory cells in endocrine glands produce hormones, which have specific effects on the body. The production of these hormones can be controlled by other hormones or by the nervous system. Since the hormones are carried to the appropriate locations by the blood stream,

this route can be used as a feedback system to stimulate a gland to increase or decrease production of a hormone.

For example, one hormone produced in the pituitary gland is thyroid-stimulating hormone (TSH). When TSH reaches the thyroid via the blood stream, it stimulates the thyroid to produce its hormone thyroxin. Thyroxin is added to the blood stream and eventually reaches the pituitary. The increase in the thyroxin level stimulates the pituitary to slow down or stop producing TSH. When the circulating level of thyroxin falls below a maintenance level, the pituitary will again be stimulated to produce TSH.

This system is called negative feedback. A hormone of one endocrine gland stimulates another to produce hormone. The production of the second hormone has an inhibiting effect on the secretion of the first gland. This negative feedback loop occurs in several of the endocrine glands.

(a) The pituitary also secretes a hormone called adrenocorticotropic hormone (ACTH) which creates a negative feedback loop with hormones produced in the adrenal cortex. Label glands A and substances B and C in the diagram of the loop on the right.

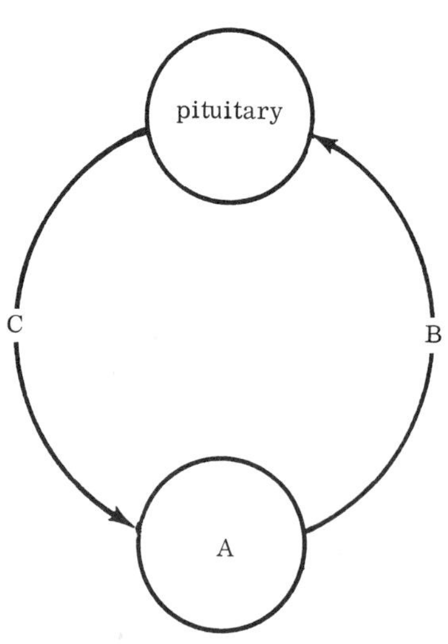

(b) What would be the effect of secretion of hormones by the cortex of the adrenal glands on ACTH secretion in the pituitary? ___________

____

(c) The interaction between follicle-stimulating hormone (FSH) from the pituitary and the hormone estrogen secreted by the ovary is much the same as that of TSH and thyroxin or ACTH and adrenal cortex hormones. Explain the relationship between FSH and estrogen.

____

____

- - - - - - - - - - - - - - - - - -

(a) A—adrenal cortex; B—hormones; C—ACTH
(b) decrease ACTH secretion (slow it down)
(c) increase in FSH produces increase in estrogen which results in decrease in FSH (typical negative feedback inhibition loop)

4. Negative feedback is a relatively slow mechanism for controlling hormonal secretion. Direct nerve stimulation from the autonomic nervous system provides very quick control of secretions in the adrenal medulla (outer part) and posterior parts of the pituitary. Some nerve cells in the hypothalamus produce chemicals that directly affect secretions in the anterior pituitary. In negative feedback, the blood level of hormones determines secretion of controlling hormones. Blood levels of other substances, such as sugar or calcium, control the secretions of the endocrine pancreas and parathyroid glands in a more direct type of feedback.

(a) Two types of endocrine glands are stimulated directly by the nervous system. Name them and tell where each type is located. __________

____

(b) Two types of endocrine glands are found in the neck. Name them and state the means of control that stimulates each. ______________

____

____

(c) Chemicals produced in nerve cell bodies of the hypothalamus influence an endocrine gland very near the brain. Which one? ________

____

(d) Name three glands in your body that are controlled by a negative feedback technique. ______________________________________

- - - - - - - - - - - - - - - - - -

(a) adrenal medulla in abdominal cavity; posterior part of pituitary, near brain
(b) thyroid—blood level of hormones (negative feedback); parathyroid—blood calcium level
(c) pituitary (anterior lobe)
(d) thyroid, adrenal cortex, and testis or ovary

The pituitary gland

5. Because of its major effects on four other endocrine glands, the pituitary gland warrants a detailed investigation. The pituitary gland has been classically divided into two parts called the anterior lobe and posterior lobe. In recent years, this gland has been intensively studied, resulting in a new name, the hypophysis, and a division into the adenohypophysis and neurohypophysis. The adenohypophysis includes the anterior lobe, while the neurohypophysis includes the posterior lobe. Each division also includes other minor components.

   The hypophysis is directly attached to the brain; the neurohypophysis is a direct extension of the hypothalamus. Thus the entire hypophysis is covered by the meninges of the central nervous system. It lies in a depression in the sphenoid bone, just a little posterior to the optic chiasma. The hypophysis is about the size of a pea, and is attached to the hypothalamus by a stalk extending from the neurohypophysis.

   (a) A more recent name for the pituitary gland is the ______________.

   (b) What division of the hypophysis includes the following?

   anterior lobe of pituitary ____________________

   posterior lobe of pituitary ____________________

   (c) To what portion of the brain stem is the hypophysis connected?

   ____

- - - - - - - - - - - - - - - - - - -

(a) hypophysis
(b) adenohypophysis; neurohypophysis
(c) hypothalamus

6. The posterior lobe or neurohypophysis is directly connected to the hypothalamus. The anterior lobe or adenohypophysis is connected to the hypothalamus by a network of capillaries called the pituitary portal system (or, if you prefer, the hypothalamico-hypophyseal portal system). Chemical substances produced in the hypothalamus can be released into the blood of this portal system and transported directly to the anterior lobe, there to affect the production of hormones by the adenohypophysis.

   Several hormones are produced in the anterior lobe. We noted earlier that some of these were influenced by negative feedback from other endocrine glands. Thyroid-stimulating hormone (TSH) controls the production and release of thyroid hormones. Adrenocorticotropic hormone (ACTH) controls secretory activity of the adrenal cortex. Follicle-stimulating hormone (FSH) causes ova or sperm (human sex cells) to mature and sex hormones to be released. Production of a fourth hormone is also influenced by negative feedback. This is called luteinizing hormone (LH) in women and interstitial cell stimulating hormone (ICSH) in men. LH or ICSH stimulates the production of sex hormones: progesterone in the ovary or testosterone in the testis. The negative feedback in all these is directed toward the hypothalamus. Chemical regulatory factors are then released by the hypothalamus into the pituitary portal system from where they influence secretion.

(a) What connects the adenohypophysis to the hypothalamus? ______________________________

(b) Indicate the effect of each of the following anterior lobe hormones.

TSH ______________________________

FSH ______________________________

ACTH ______________________________

LH ______________________________

ICSH ______________________________

(c) What process is diagrammed on the right? ______________________________

Give names to:

A ______________

B ______________

C ______________

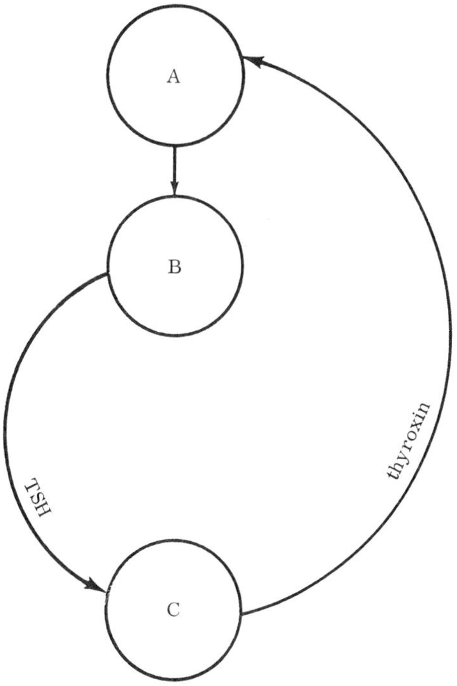

- - - - - - - - - - - - - - - - - -

(a) a capillary network called the pituitary portal system
(b) TSH—controls production and release of thyroid hormones

FSH—causes maturation of ova or sperm
ACTH—controls secretion of adrenal cortex hormones
LH—production of female sex hormone (progesterone)
ICSH—production of male sex hormone (testosterone)

(c) negative feedback
A—hypothalamus
B—hypophysis (pituitary), more specifically the adenohypophysis or anterior pituitary
C—thyroid

7. Another critical hormone produced in the adenohypophysis is somatotrophic hormone (STH). This substance controls body growth as well as influencing fat and sugar metabolism; hence STH is often called growth hormone. Over or under secretion of somatotrophic hormone during childhood results in giantism or dwarfism. Prolactin, a hormone which sustains lactation (milk production) after pregnancy, completes the roster of anterior lobe hormones. Both somatotrophic hormone and prolactin production are controlled by regulatory factors from the hypothalamus, although no negative feedback has been identified.

Identify the anterior pituitary hormone that has each of the following effects.

(a) controls maturation of ova and sperm ______________

(b) sustains milk production after childbirth ______________

(c) influences overall body size ______________

(d) regulates release of thyroid hormones ______________

(e) controls hormone production by adrenal cortex ______________

(f) control sex hormone production (two) ______________

- - - - - - - - - - - - - - - - - -

(a) FSH (follicle stimulating hormone)
(b) prolactin
(c) STH (somatotrophic hormone or growth hormone)
(d) TSH (thyroid stimulating hormone)
(e) ACTH (adrenocorticotrophic hormone)
(f) LH or ICSH (luteinizing hormone, interstitial cell stimulating hormone); FSH also has effects

8. The posterior lobe in the neurohypophysis does not produce hormones, but it does store two hormones that are produced in neuron cell bodies in the hypothalamus. The product travels down the axons in the connecting stalk to be stored in the terminals in the neurohypophysis. Oxytocin

stimulates contractions of the uterus during and after childbirth, and is involved in milk production. Antidiuretic hormone (ADH) produces a more concentrated urine by stimulating the kidney to remove water from urine and replace it in the blood.

(a) How is release of hormones from the neurohypophysis stimulated?

____

(b) What stimulates release of hormones from the adenohypophysis?

____

(c) By what type of cell are ADH and oxytocin produced? __________

(d) In what structure are they produced? __________

(e) In what structure are they stored? __________

(f) Into what are they released? __________

- - - - - - - - - - - - - - - - - -

(a) electrical impulses from hypothalamus; (b) a blood-carried regulatory chemical; (c) neuron; (d) hypothalamus; (e) neurohypophysis (posterior pituitary); (f) into capillaries, that is, directly into the bloodstream

9. Complete the following chart for a summary of hormones released by the hypophysis.

Hormone	Where produced	Responds to what type of stimulation	Affects what?
STH			
TSH			
ACTH			
FSH			
LH (ICSH)			
Prolactin			
Oxytocin			
ADH			

- - - - - - - - - - - - - - - - - -

STH	Adenohypophysis	chemical	body size and metabolism
TSH	Adenohypophysis	chemical and negative feedback	thyroid function
ACTH	Adenohypophysis	chemical and negative feedback	adrenal cortex function
FSH	Adenohypophysis	chemical and negative feedback	sperm or ovum maturation
LH (ICSH)	Adenohypophysis	chemical and negative feedback	sex hormone production
Prolactin	Adenohypophysis	chemical	milk production
oxytocin	Hypothalamus	electrical (neural)	uterine contractions
ADH	Hypothalamus	electrical (neural)	absorption of water from urine

The thyroid

10. The thyroid gland is located in the anterior neck region, just above the sternoclavicular joint. A lobe is found on the right and left, connected by a narrow isthmus of tissue across the front of the trachea, or windpipe. As with other endocrine glands, the blood supply to the thyroid is abundant.

    The functional unit of the thyroid gland is the thyroid follicle, a hollow sphere of cells. The follicle is lined with epithelial cells that produce the hormone thyroxin. Thyroxin is then stored in a colloid form (similar to jello) in the center of the thyroid follicle until its release is stimulated by blood-borne TSH. Thyroxin controls most catabolic metabolism, or the speed and efficiency with which the body functions.

    Another thyroid hormone, calcitonin, is produced by parafollicular cells (para means near) located between the thyroid follicles. Calcitonin production is controlled by blood calcium levels, and functions in lowering the levels of calcium and phosphate ions in the blood.

    (a) Name two hormones produced by the thyroid gland. ______________

    ____

    (b) What is the function of the hormone produced by the thyroid follicle?

    ____

    (c) What mechanism controls the production of the hormone calcitonin by the parafollicular cells? ______________________________

(d) What sort of interaction might occur between the hypothalamus and the thyroid gland? ______________________________

- - - - - - - - - - - - - - - - - - -

(a) thyroxin and calcitonin; (b) controls catabolic metabolism; (c) level of calcium in the blood; (d) negative feedback inhibition between TSH and thyroxin

11. Thyroid gland difficulties are generally concerned with thyroxin production. Hypersecretion (too much) results in an increase in activity level, an increased heart rate, and in some cases protruding eyeballs. Hyposecretion (too little) is reflected in lowered metabolic rate with increased fatigue, slow heart rate, low blood pressure, and low body temperature.

    When hypothyroidism begins before birth, a child may be retarded mentally and dwarfed physically; this condition is called cretinism. Hypothyroid states after birth are called myxedema.

    Thyroxin consists, to a large extent, of iodine. The widespread use of iodized table salt has done much to keep hypothyroidism under control.

    (a) A person who always seems agitated, keeps turning down the thermostat, and has a very fast pulse may have a thyroid disorder.

    Which hormone would be involved? ______________________

    Would hypersecretion or hyposecretion be the problem? __________

    ____

    (b) Identify each of the symptoms below with hyperthyroidism or hypo-hypothyroidism.

    protruding eyes ______________________

    low blood pressure ______________________

    excessive fatigue ______________________

    myxedema ______________________

    (c) Identify the hormone associated with each of the following.

    phosphates ______________________

    iodine ______________________

    calcium ______________________

- - - - - - - - - - - - - - - - - - -

(a) thyroxin, hypersecretion
(b) hyperthyroidism
hypothyroidism
hypothyroidism
hyperthyroidism
(c) calcitonin
thyroxin
calcitonin

12. The four parathyroid glands are located on the posterior aspect of the thyroid, two on each lobe. The cells of the parathyroid produce parathormone (PTH), which works in cooperation with calcitonin to control calcium and phosphate usage by the body.

One critical way this is done is in the control of resorption or breakdown of bone. A fall in blood calcium level signals the parathyroids to increase production of parathormone. A rise in blood calcium level signals the onset of calcitonin production. A deficiency of parathormone results in low blood calcium levels. This causes severe muscle cramps, and generally disrupts muscle functioning. The total removal of all four parathyroid glands can result in bone destruction.

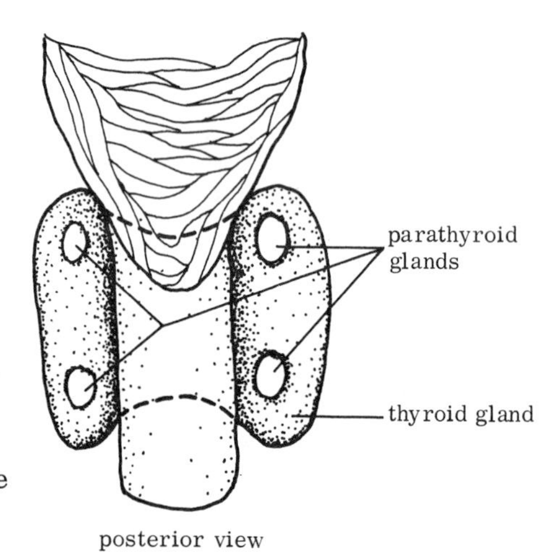

posterior view

(a) Are the parathyroid glands more closely interrelated <u>functionally</u> to the hypothysis or the thyroid? ______________________

(b) How are PTH and calcitonin stimulated to control calcium and phosphate metabolism? ______________________
____

(c) What would be the effect of thyroid destruction (but not parathyroid) on the blood calcium levels? ______________________
____

- - - - - - - - - - - - - - - - - -

(a) thyroid (physically and functionally)

(b) calcitonin production is stimulated by a rise in blood calcium, while PTH production begins when blood calcium level falls
(c) probably bone resorption would occur as body would continue to increase the blood calcium levels

The adrenal glands

13. The two adrenal glands are located in the back, at about the level of the tenth rib. One gland caps each kidney. The outer part of each adrenal gland is the cortex, while the medulla is in the center, completely surrounded by cortical tissue.

The adrenal medulla produces two relatively similar hormones, epinephrine and norepinephrine. These chemicals have the same general effect as stimulation of the sympathetic nervous system; they prepare the body to cope with stress or a crisis.

The adrenal cortex produces hormones that differ in chemical structure from those already studied. The cortical hormones are not protein in nature, but are steroids, the most important of which are glucocorticoids, such as cortisone. These chemicals have several functions, among which are to increase metabolism of carbohydrates and proteins.

(a) Identify the two regions in the adrenal gland. ____________________

____

(b) Name one hormone produced in each region. ____________________

____

(c) Name three specific effects of increased epinephrine secretions.

____

____

- - - - - - - - - - - - - - - - - -

(a) cortex and medulla
(b) cortisone (or glucocorticoids) in cortex; epinephrine or norepinephrine in medulla
(c) increase heart rate, breathing, and blood pressure (or any of the sympathetic effects discussed in Chapter 5)

14. Secretion of hormones by the adrenal cortex is controlled primarily by ACTH secretion by the hypophysis. A deficiency of cortical hormone secretion results in Addison's disease, which is characterized by anemia, weakness, digestive problems, and low blood sodium levels. Cushing's disease results from hypersecretion of glucocorticoids, with such symptoms as a moon-shaped face and fat concentrated on the torso with a pendulous abdomen.

Nerve fibers from the sympathetic nervous system stimulate release of epinephrine and norepinephrine by the adrenal medulla. Disorders of secretion of medullary hormones are quite rare.

(a) Excess secretion of ACTH might result in which disease? ________

____

(b) What would be the effect of a slight decrease in glucocorticoids in the blood stream? ________________________ What about a severe decrease? ________________________

(c) What mechanism controls the secretion of medullary hormones?

____

(d) The adrenal gland has two secretory portions, the cortex and the medulla. Are these under the same type of stimulus control?

____

- - - - - - - - - - - - - - - - - -

(a) Cushing's disease
(b) increased ACTH release; Addison's disease
(c) direct nerve stimulation from the sympathetic nervous system
(d) no (cortex is negative feedback, medulla is direct nerve stimulation)

The pancreas

15. The pancreas is located in the abdominal cavity, just below the stomach. It functions as both an exocrine digestive gland and as an endocrine gland. In Chapter 11 we will consider its digestive function.

The endocrine cells of the pancreas are located in many small groups called islets of Langerhans. The cells in each islet produce two hormones called glucagon and insulin. Both affect levels of sugar in the blood, with glucagon raising the blood glucose level and insulin lowering it.

The blood glucose level controls secretion of the hormones; a fall in blood glucose causes glucagon production while a rise causes insulin secretion. When the body does not produce insulin, the condition is called diabetes mellitus. While diabetes is basically a disorder in the use of carbohydrates from food, it causes problems in most of the body systems if unchecked. Daily administration of insulin with concurrent attention to diet can control the condition to keep the body functioning adequately.

(a) In what areas of the pancreas are endocrine cells found? ________

____

(b) Name the two hormones produced in the pancreas and contrast their control mechanisms. ______________________________

____

(c) What other two endocrine hormones are interrelated in much the same way as the pancreatic hormones? ______________________

What controls their production? ______________________

- - - - - - - - - - - - - - - - - -

(a) islets of Langerhans
(b) glucagon and insulin; glucagon is produced when blood sugar falls, insulin when it rises
(c) parathormone and calcitonin; blood calcium levels

The gonads, testes in the male and ovaries in the female, are treated in detail in Chapters 14 and 15. We indicated that FSH and LH (or ICSH) are involved in the control of sex hormones. These are fairly standard negative feedback systems. Too much or too little secretion can result in variations of secondary sex characteristics (hair or fat distribution, voice) in either sex, or development of the signs of the opposite sex.

16. Complete the following chart.

Gland	Functional division	Hormones produced	What controls or stimulates production	Effect of hormone
Thyroid	________	________	________	________
		________	________	________
Parathyroid	________	________	________	________
Adrenal	________	________	________	________
	________	________	________	________
Pancreas	________	________	________	________
		________	________	________
Gonads	________		________	
	________		________	

- - - - - - - - - - - - - - - - - - -

Thyroid	none	thyroxin calcitonin	TSH blood calcium	Control metabolism Lowers calcium in blood
Parathyroid	none	PTH	blood calcium	Raises calcium in blood
Adrenal	cortex	glucocor-ticoids	ACTH	Carbohydrate and protein metabolism
	medulla	epinephrine	Sympathetic nerves	Prepares for stress
Pancreas	Islets of Langer-hans	glucagon	blood sugar	Raises blood sugar
		insulin	blood sugar	Lowers blood sugar
Gonads	ovary testis		FSH, LH FSH, ICSH	

SELF-TEST

This Self-Test is designed to show how well you have mastered this chapter's objectives. Answer each question to the best of your ability. Correct answers and review instructions are given at the end of the test.

1. What distinguishes endocrine glands from exocrine glands? ________

____

2. (a) What mechanism regulates the production of thyroxin? ________

____

   (b) What is the function of thyroxin? ________________

3. For each hormone below, indicate whether it is produced in the adenohypophysis, neurohypophysis, or hypothalamus.

   (a) follicle stimulating hormone ________________

   (b) oxytocin ________________

   (c) antidiuretic hormone ________________

   (d) prolactin ________________

4. What is the primary purpose of the pituitary portal system? ________

____

5. In a condition called acromegaly, certain bones and joints enlarge, resulting in a distorted appearance. This condition only occurs in adults.

   What hormonal state might cause it? ________________

____

6. What effect do thyroid hormones have on parathyroid hormone production? ________________________________

7. How is the production of pancreatic hormones regulated? ________

____

8. What controls the production of hormones from the adrenal medulla?

____

9. What endocrine glands are controlled neurally? ________________

10. Which endocrine glands are directly affected by pituitary secretions?

____

Answers

Compare your answers to the Self-Test questions with those answers given below. If all of your answers are correct, you are ready to go on to the next chapter. If you missed any, review the frames indicated in parentheses following the answers. If you missed several questions, you should probably reread the entire chapter carefully.

1. Endocrine glands release hormones directly into the blood stream, rather than through a system of ducts, as in exocrine glands. (frame 1)
2. (a) negative feedback to pituitary for thyroid stimulating hormone (TSH) production (frame 3)
   (b) maintain body metabolism (frame 10)
3. (a) adenohypophysis (frame 6)
   (b) hypothalamus (stored in neurohypophysis) (frame 8)
   (c) hypothalamus (stored in neurohypophysis) (frame 8)
   (d) adenohypophysis (frame 7)
4. carry feedback messages (regulatory factors) to adenohypophysis (frame 6)
5. oversecretion of STH or growth hormone (in adulthood) (frame 7)
6. Calcitonin works in cooperation with parathormone (PTH) in response to blood calcium levels. (frame 12)
7. Insulin and glucagon are secreted in response to a rise or fall, respectively, in blood sugar levels. (frame 15)
8. sympathetic nervous system (frame 13)
9. pituitary and adrenal glands (frame 14)
10. thyroid adrenal (cortex), ovaries, and testes (frame 9)

PART IV

Maintenance Systems of the Body

We have seen how the human body is supported and moved by the skeletal and muscular systems. We have seen how parts of the body communicate and influence other parts through the nervous and endocrine systems. In this unit we will examine the maintenance systems of the body.

The human body interacts regularly with the world around it in order to maintain itself, or keep itself going. To survive as a successful organism, the body must receive and incorporate nourishment and oxygen from the surrounding environment via its digestive and respiratory systems. The circulatory and lymphatic systems transport the nourishment and oxygen to all the cells of the body. The excretory organs remove waste products from the body, while the skin encases the organism and protects it from undesirable elements.

Working together, the six maintenance systems enable the human organism to thrive and live a healthy individual life.

CHAPTER EIGHT
The Circulatory System

The previous unit dealt with communication systems in the body, including the nervous and endocrine systems. In this unit on maintenance systems, it seems fitting that we begin with the circulatory system, which performs a small portion of its maintenance tasks by transporting communication products of the endocrine system. After you complete your study of this chapter, you will be able to:

- specify the primary function of each of the three components of the circulatory system;
- describe or identify elements contained in blood, including erythrocytes, granular and nongranular leukocytes, platelets, and plasma;
- differentiate between descriptions of the microscopic structure of arteries, veins, arterioles, venules, and capillaries;
- trace the flow of blood through the heart;
- identify major structural elements in the heart, including names of chambers, valves, and vessels connecting to it;
- compare the muscle in the heart wall to that found in blood vessel walls;
- compare the pulmonary and systemic circulations.

BASIC COMPONENTS OF BLOOD

1. Any circulating system needs three basic components: a fluid to circulate, tubes or channels to form a pathway, and a pump to keep the fluid moving. The three basic components of the human circulatory system are blood, blood vessels, and the heart. The blood transports many different elements as it circulates within vessels, including hormones secreted by various endocrine glands, oxygen from and carbon dioxide to the lungs, and nutrients from the digestive tract to all the cells of the body. Its primary function is homeostasis—maintaining a balance in the internal environment of the body.

(a) Write in the name of the component of the blood circulatory system that plays each role listed below.

Fluid to circulate ____________________

Tubes to contain the fluid ____________________

Pump to continue circulation ____________________

(b) Which of the situations below describes the blood serving its function of homeostasis? ____________________

(1) The effect of secretion of thyroid stimulating hormone by the pituitary.
(2) The sensation of vision.
(3) Breathing.

(c) Hormones, gases, and nutrients, as well as minerals, salts, and water, all contribute to the blood circulatory system's primary function of maintaining ____________________.

(d) The blood itself contributes most directly to maintaining a uniform environment for the cells of the body. The heart contributes to this homeostasis by serving as a ____________________. The conduits for the circulation are provided by the ____________________.

- - - - - - - - - - - - - - - - - - -

(a) blood; blood vessels; heart
(b) 1 and 3
(c) homeostasis
(d) pump; blood vessels

2. Blood is not, as may appear, a simple liquid. It is a tissue, made up of various elements, each with distinguishing characteristics and functions.
   Blood, as a tissue, consists of cells, or cell-like structures, with much intercellular material. Based on this fact alone, you should be able to classify blood into one of the four tissue types discussed in Chapter 2. Which is it? ____________________

- - - - - - - - - - - - - - - - - - -

connective tissue

3. Blood is a specialized type of connective tissue that contains no fibers. It does contain "formed elements" that float in a liquid called plasma. Thus, the formed elements correspond to matrix or ground substance.

The quantity of blood in the body is about 7 percent of the weight of the body. Thus, a pre-teenager weighing 100 pounds would have 7 pounds of blood. The plasma makes up about 55 percent of the volume of blood. The child with 7 pounds of blood would have about 3.8 pounds of plasma. Formed elements make up the balance of the blood.

(a) A man weighing 200 pounds would have approximately how much blood? ____________________

(b) What percentage of the blood is made up of formed elements? _____ ____________________

- - - - - - - - - - - - - - - - - -

(a) 14 pounds (200 times .07)
(b) 45% (varies, of course, with individual and health)

Formed elements

Formed elements can be divided into three main groups: red blood corpuscles, white blood cells, and platelets. In this next section, we will consider each of these types.

4. Red blood corpuscles are sometimes called red blood cells, but their technical name is erythrocyte (erythro means red). The abbreviation RBC is also widely used.

   The erythrocyte is not a true cell, since it has no nucleus in its mature form. In the blood stream erythrocytes are about 700 times as numerous as white blood cells, and about 25 times as numerous as platelets. The erythrocyte is shaped like a disc, concave on both sides so that it is thinner in its central section than on the edge of the disc. The cytoplasm of an erythrocyte is almost filled with an iron-containing material called hemoglobin, which gives a pinkish color to each corpuscle and the red color to blood.

   (a) Which type of formed element is most numerous in blood—red blood corpuscle, white blood cell, or platelet? ____________________

   (b) What gives the pinkish color to an erythrocyte? ________________

   (c) Which of the shapes below is similar to that of an erythrocyte? ____________________

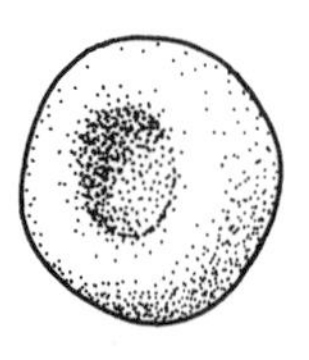

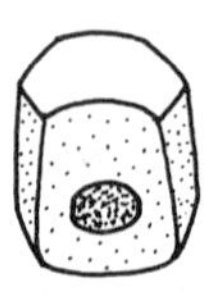

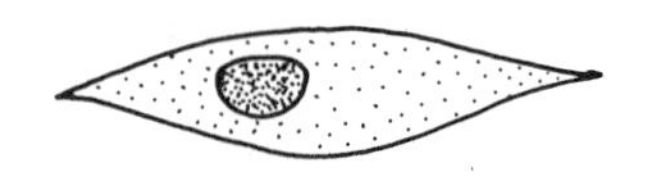

- - - - - - - - - - - - - - - - - -

(a) red blood corpuscle; (b) hemoglobin or iron; (c) the one on the left, a biconcave disc

5. The erythrocyte must be flexible enough to pass through very tiny blood vessels in the body and yet return to the basic disc shape which furnishes a large surface area. The only organelle found in the erythrocyte is the microtubule. Forming a ring around the outer margin of the corpuscle, the microtubules are thought to aid in maintaining the double-concave surface that exposes a large amount of the hemoglobin so that it can more easily take up and release oxygen and carbon dioxide. Transporting oxygen and carbon dioxide between the lungs and the cells of the body is the function of the erythrocytes.

   (a) Which would you expect to see in a high magnification view of an erythrocyte—mitochondria, microtubules, or nucleus? ______________________________

   (b) Is an erythrocyte always the same shape as it flows through the body? ____________________

   (c) What is contained in the erythrocyte that combines easily with oxygen and carbon dioxide? ____________________

- - - - - - - - - - - - - - - - - -

(a) microtubules
(b) no; it flexes easily when necessary, but always returns to the disc-shape
(c) hemoglobin (more about this in Chapter 10)

6. White blood cells, also called white corpuscles or leukocytes (leuko means white) are more diverse than are erythrocytes; that is, they have several different shapes and colors. Basically they are divided into two groups—granular and nongranular.

   The granular leukocytes contain granules within their cytoplasm; there are three types which stain different colors. The nuclei of these granular leukocytes are lobed and segmented, which also helps to identify them and distinguish them from the nongranular leukocytes which have a more regularly shaped nucleus.

   (a) Would a formed element with an oval nucleus be an erythrocyte, a granular leukocyte, or a nongranular leukocyte? ______________________________

(b) How would you classify a formed element with large granules and a nucleus with three lobes? ______________________

(c) What type of formed element has no nucleus or granules? ____________ ______________________

- - - - - - - - - - - - - - - - - -

(a) nongranular leukocyte; (b) granular leukocyte; (c) erythrocyte

7. Normally, most leukocytes in blood are granular. Three types of these are seen. The most common leukocyte of all, the neutrophil, makes up 60 percent of the white cells. Its granules stain lightly pink and blue with a neutral dye (giving it the name neutrophil). The neutrophil nucleus has from three to five lobes.

   Another granular leukocyte is the eosinophil, which has large reddish-orange granules that stain with an eosin dye. Only about 1 to 3 percent of the leukocytes are eosinophils, and their nuclei have fewer lobes than do the neutrophil.

   The third granular leukocyte is the basophil, whose very large, deep purple granules take up the basic dye. The basophil nucleus is sometimes S-shaped, but is usually covered up with the large purple granules. The basophils make up only about 1 percent of the leukocytes, and are the smallest in overall size. They are just a bit larger than erythrocytes, while the neutrophil and eosinophil are about twice the size of an erythrocyte.

   (a) What granular leukocyte is most common? ______________________

   (b) What type of leukocyte has dark purple granules? ______________

   (c) What type of leukocyte has reddish-orange granules? ____________

   (d) Which would be smaller—an eosinophil, a basophil, or an erythrocyte? ______________________

- - - - - - - - - - - - - - - - - -

(a) neutrophil; (b) basophil; (c) eosinophil; (d) erythrocyte

8. The nongranular leukocytes consist of two types. Lymphocytes make up about 30 percent of the leukocytes and monocytes make up about 5 percent. Lymphocytes are just a bit larger than a basophil, with a nucleus which takes up almost all the space inside the cell. Very little cytoplasm is present in a lymphocyte.

   The monocytes are larger than the neutrophils and eosinophils.

Monocytes also have large nuclei, but it takes up only about half the cell area. Thus, more cytoplasm would be seen in a monocyte than in a lymphocyte. The large nucleus in a nongranular leukocyte is not lobed; it may be indented in a monocyte but is generally round or oval in shape.

(a) Which is the more numerous nongranular leukocyte? ____________

(b) Which nongranular leukocyte has less cytoplasm in proportion to its nucleus? ____________________

(c) Which leukocyte is most likely to have an indented nucleus? ______ ____________________

(d) Which is the largest of the leukocytes? ________________________

- - - - - - - - - - - - - - - - - -

(a) lymphocyte; (b) lymphocyte; (c) monocyte; (d) monocyte

9. All of the leukocytes function in protection and defense of the body. All are capable of phagocytosis, or capturing and digesting foreign material. Leukocytes are the body's defenders; the blood stream delivers them to any site where the skin may be broken or where some internal condition sets up a signal. The neutrophils are usually the first to attack bacteria.

   Most leukocytes live for only a few days, but some lymphocytes are now thought to be more durable and to play a critical role in the immunity system of the body.

   (a) Which cells are the first to attack bacteria? ____________________

   (b) Which cells are most involved with immunity? __________________

   (c) What is the process of trapping and digesting foreign material called? ____________________

- - - - - - - - - - - - - - - - - -

(a) neutrophils; (b) lymphocytes; (c) phagocytosis

10. Platelets, or thrombocytes, make up the last group of formed elements. Platelets are small (about one-fourth the size of an erythrocyte), contain no nucleus, and are irregular in shape. Actually, platelets are fragments of cytoplasm of very large cells (megakaryocytes) located in the bone marrow. Platelets contain several critical blood clotting elements, and they even tend to stick together in clumps as they float in the plasma.

    (a) Name two formed elements in blood that have no nucleus. ________ ________________________________

(b) Name two formed elements that do have a nucleus. ______________

____

(c) Which of the nonnucleated formed elements is smaller? __________

____

(d) What do platelets contribute to blood? __________________________

- - - - - - - - - - - - - - - - - -

(a) erythrocyte, platelet
(b) (any two) neutrophil, eosinophil, basophil, lymphocyte, monocyte, granular and nongranular leukocyte
(c) platelet
(d) clotting elements

11. In the chart below, indicate whether a nucleus is present or has lobes. Write in the color of granules in the granular leukocytes. Then enter briefly the function of each.

	Nucleus	Granules	Function
Erythrocyte	______________	______________	______________
Neutrophil	______________	______________	______________
Eosinophil	______________	______________	______________
Basophil	______________	______________	______________
Lymphocyte	______________	______________	______________
Monocyte	______________	______________	______________
Platelet	______________	______________	______________

- - - - - - - - - - - - - - - - - -

none	none	carry oxygen and carbon dioxide
lobes	pink, blue	phagocytosis or defense
lobes	red, orange	phagocytosis or defense
lobes	purple	phagocytosis or defense
present	none	immunity
present	none	defense
none	none	clotting

12. Rank the formed elements below in terms of size (largest to smallest) and quantity (most numerous to least numerous). The largest and most numerous elements have been filled in.

Size		Quantity
______	Erythrocyte	1
______	Neutrophil	______
______	Eosinophil	______
______	Basophil	______
______	Lymphocyte	______
1	Monocyte	______
______	Platelet	

- - - - - - - - - - - - - - - - - -

6	1
2 or 3	2
2 or 3	5
5	6
4	3
1	4
7	

13. The formed elements of the blood float in the plasma, which is the liquid portion. Over 90 percent of the plasma is water. Various proteins, including several more clotting factors, make up another 6 to 8 percent of the plasma. The remainder is salts, dissolved gases, wastes, food, and hormones.

The plasma's function is to transport; it moves the erythrocytes from lung to cell with their cargo of oxygen and from cell to lung with carbon dioxide. It transports hormones from secretor to target organ. It carries platelets and proteins to sites where clotting is required, and carries defense cells to points where bacterial invaders may be.

(a) Do you recall what percent of the blood is plasma? ______________

(b) Where are the clotting factors in the blood localized? ______________

____

(c) If you were to divide blood into two constituents, what would they be? ______________________________

- - - - - - - - - - - - - - - - - -

(a) 55 (frame 3); (b) platelets and plasma proteins; (c) formed elements and plasma

Blood review

14. Identify the elements in blood from the following description.
    (a) a formed element with no nucleus and a biconcave disc shape

    ____

    (b) contains over 90 percent water ____________________
    (c) functions in immunity ____________________
    (d) large round nucleus; very little cytoplasm ____________________
    (e) large purple granules ____________________
    (f) light staining granules; lobed nucleus ____________________
    (g) contains hemoglobin ____________________
    (h) one-fourth the size of erythrocyte ____________________

- - - - - - - - - - - - - - - - - - -

(a) erythrocyte; (b) plasma; (c) lymphocyte; (d) lymphocyte; (e) basophil; (f) neutrophil; (g) erythrocyte; (h) platelet

BLOOD VESSELS

Blood circulates in a closed system made up of blood vessels. The blood is forced by the heart pump to course repeatedly through the body, to bring needed elements to every cell, following the route laid down by blood vessels. We will consider first the capillaries, the microscopic tubes which are in contact with all the cells and tissues. Then we will consider the arteries, which carry blood away from the heart toward the capillaries, and finally the veins, which return blood to the heart.

15. Capillaries are small tubes whose walls are very thin—the thickness of just one very thin endothelial cell. The diameter of the tube is just about the size of one erythrocyte. There are literally miles of capillaries in the body. The very small size of each, combined with the very large number, gives a tremendous surface area for erythrocytes to exchange oxygen for carbon dioxide, for nutrients in the plasma to be passed through the very thin endothelial cytoplasm, or for leukocytes to leave the blood stream to fight invaders.
    (a) What is the approximate diameter of a capillary? ____________________

    ____

(b) Would you expect to find capillaries in the skin? ________________

(c) Would you be more likely to find a capillary returning blood to the heart or in your left hand? ______________________

- - - - - - - - - - - - - - - - - -

(a) about the size of an erythrocyte (7 to 9 microns); (b) yes—but not in the epithelial part; (c) left hand

16. The capillary wall is only one cell thick. The walls of small arteries, called arterioles, that lead directly into the capillaries, are similar, but do contain some smooth muscle arranged circularly as a second, outer layer. Arterioles further from the capillaries contain a layer of connective tissue as the outside coating. These very small arteries exert control on the blood pressure of the body by means of their circular layer of smooth muscle.

    (a) Which has more smooth muscle—a capillary or an arteriole?

    ____

    (b) Which has a single cell forming its wall—a capillary or an arteriole?

    ____

    (c) Which would lead directly into a capillary—a tube with an outer layer of smooth muscle or one with an outer layer of connective tissue?

    ____

- - - - - - - - - - - - - - - - - -

(a) arteriole; (b) capillary; (c) smooth muscle

17. Most arteries have three layers in their walls. A thick outer connective tissue layer is called the tunica adventitia. The middle layer, called tunica media, is a combination of fibroelastic connective tissue and smooth muscle fibers. The combination varies with the size of the artery. Larger arteries, such as the aorta, contain more elastic fibers, medium ones contain more smooth muscle fibers than elastic tissue, while small arteries contain only smooth muscle in their tunica media. The inner layer, tunica intima, consists of the endothelial lining with a very thin connective tissue membrane.

    (a) What are the three layers of an artery wall? ____________________

    ____

(b) An artery has a thick outer layer of connective tissue, a middle layer of mostly elastic fibers with a few smooth muscle cells. Is this artery large, medium, or small? ____________________

(c) Which size artery has only smooth muscle cells in its tunica media?

____

- - - - - - - - - - - - - - - - - -

(a) tunica adventitia; tunica media; tunica intima
(b) large (medium would have mostly smooth muscle, less elastic)
(c) small (arteriole)

18. Blood flows from the heart into large arteries, then down the arterial tree into medium arteries and on into arterioles and capillaries. Blood is returned to the heart through a system of veins, beginning with small venules that parallel the arterioles.

Venules often contain just two layers, the endothelial lining surrounded by some connective tissue. Most veins, however, like most arteries, contain three layers in their walls. The adventitia in veins is very thick, and may contain longitudinally oriented smooth muscle, while the media is thin, containing little muscle or elastic fibers. Since veins do not have to withstand pressure, contractility and elasticity are not critical. The tunica intima is similar in all blood vessels.

(a) What would be the thickest layer in the wall of a medium sized vein?

____

(b) Which would contain more muscle, a medium artery or a large vein?

____

(c) Suppose you have under a microscope an example of a capillary with a vessel adjacent. How could you tell if the vessel were an arteriole or a venule? ____________________

____

- - - - - - - - - - - - - - - - - -

(a) adventitia
(b) medium artery
(c) An arteriole would have smooth muscle as its second layer, while a venule would have connective tissue.

Blood vessel review

19. Identify each of the blood vessel types described below.

    (a) connective tissue adventitia; many elastic fibers in tunica media

    ____

    (b) heavy connective tissue adventitia; very thin scattered smooth muscle in tunica media ____________________

    (c) tunica intima of endothelial cells; connective tissue second layer

    ____

    (d) single layer of endothelial cells ____________________

    (e) thick, muscular tunica media ____________________

    (f) leads directly into a capillary ____________________

    (g) leads directly from a capillary ____________________

- - - - - - - - - - - - - - - - - - - -

(a) large artery; (b) medium vein; (c) venule; (d) capillary;
(e) medium artery; (f) arteriole; (g) venule

The heart

The blood and blood vessels do not make up a complete circulatory system; the heart is needed to pump the blood through the vessels, with enough pressure so that the vital materials will be transported to the desired locations in the body.

20. The heart is located in the cavity between the lungs, the mediastinum, as indicated in the drawing. The size of the heart is approximately the size of the individual's fist. It is covered with a double-walled pericardium that separates it from the surrounding tissues.

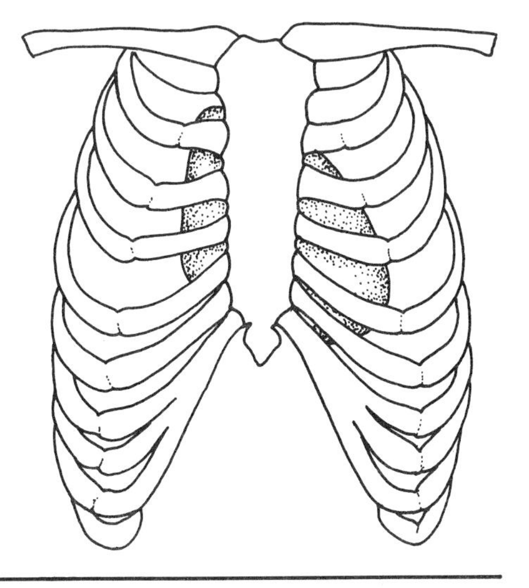

    (a) Refer to the diagram, and determine which ribs are directly in front of the heart. ____________________

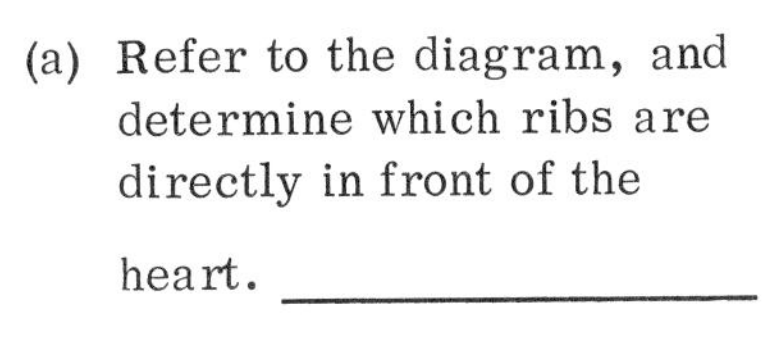

    ____

(b) Name the tissue that directly surrounds the heart. ______________

(c) Name the cavity in which the heart is found. ____________________

- - - - - - - - - - - - - - - - - - -

(a) second through sixth; (b) pericardium; (c) mediastinum

21. The heart is a muscular organ, containing four chambers as shown at the right. The chambers are divided by walls or septa (singular, septum), and connected by valves. The small receiving chambers, called atria (singular, atrium) are located in the upper part of the heart, while the two ventricles, or pumping chambers, are located in the lower part.

    The chamber labeled A, in the upper left, is the right atrium.

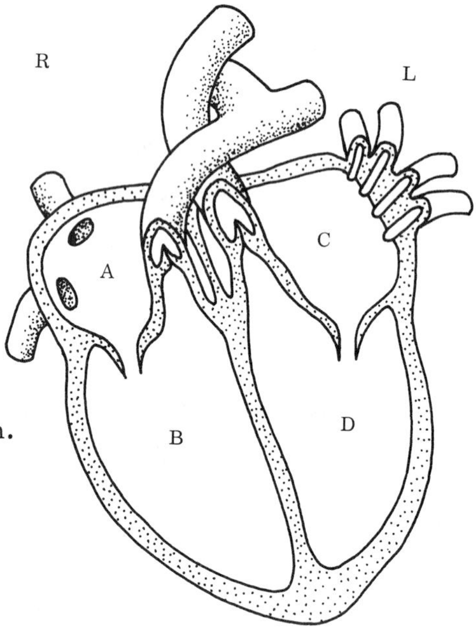

(a) Name chamber B __________

____

(b) Name chamber C __________

____

(c) Name chamber D __________

____

The wall between chambers A and B is called the right atrioventricular septum.

(d) Name the wall between chambers C and D. ____________________

The wall between chambers A and C is called the interatrial septum.

(e) Name the wall between chambers B and D. ____________________

- - - - - - - - - - - - - - - - - - -

(a) right ventricle; (b) left atrium; (c) left ventricle; (d) left atrioventricular septum; (e) interventricular septum

22. Large vessels enter both atria. Additional large vessels carry blood from the ventricles.

(a) Consider your knowledge of blood vessels. Do veins or arteries bring blood into the atria from the rest of the body? ____________

(b) What type of vessel carries blood from the ventricles toward the body cells? ______________________

- - - - - - - - - - - - - - - - - -

(a) veins; (b) artery

23. Blood enters the right atrium of the heart through the largest veins in the body, the superior and inferior vena cavae. These veins return blood from tissues in the body above and below the heart. The blood passes through the right atrioventricular valve into the right ventricle. The pulmonary artery then carries the blood away from the heart to the lungs, where it will receive a new supply of oxygen. The blood leaves the lungs to return to the heart via the pulmonary veins to the left atrium. The blood now passes through the left atrioventricular valve into the left ventricle, from where it is pumped into the aorta, the largest artery in the body. The diagram shows the relative locations of these points in or near the heart.

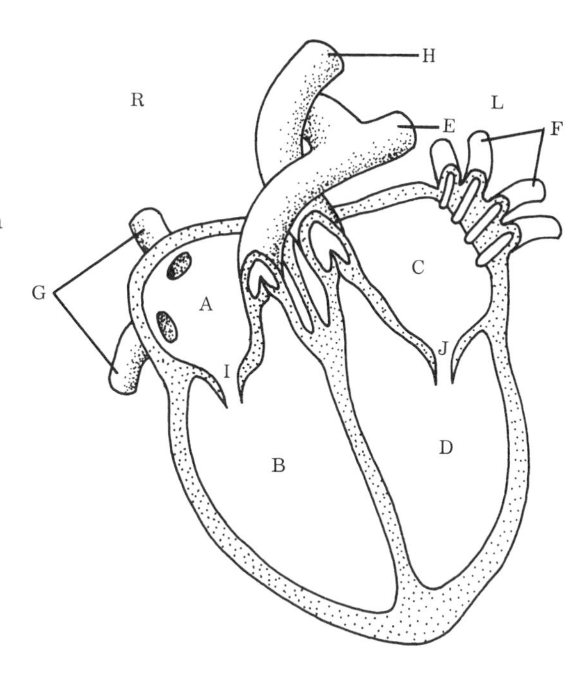

(a) Name the vessel represented by F in the diagram. ______________

(b) Name the vessel represented by G. ______________________

(c) Name the point indicated by J. ______________________

(d) Name a vessel that opens into the left ventricle. ______________

(e) Name the chambers of the heart in the order in which blood flows through them, beginning with the one to which the venae cavae return blood. ____

(f) Number the structures below in the same sequence as asked for in (e) above. The first of the series has been numbered.

_____ aorta

_____ right atrioventricular valve

_____ left atrioventricular valve

_____ pulmonary artery

_____ pulmonary vein

__1__ superior vena cava

_____ lungs

- - - - - - - - - - - - - - - - - -

(a) pulmonary vein
(b) vena cava
(c) left atrioventricular valve
(d) aorta
(e) right atrium, right ventricle, left atrium, left ventricle
(f) 7, 2, 6, 3, 5, 1, 4

24. Like the large vessels with which it is continuous, the heart wall has three layers. The walls of the heart have different names than they do in the blood vessels.

The inner layer of the heart is called endocardium; it consists of light connective tissue with an endothelial lining that gives a smooth surface and offers little resistance to blood flow.

The middle layer of the heart wall is called myocardium, and is made up of cardiac muscle, the branching striated muscle that was covered in Chapter 2. This myocardium makes up most of the mass of the heart.

The outer epicardium covers the heart with connective tissue, and continues to cover the heart as membranous pericardium. The three layers of the heart wall are continuous with the corresponding layers of blood vessel walls.

(a) Name the blood vessel wall layer that corresponds to each of the following.

endocardium _______________________

myocardium _______________________

epicardium _______________________

(b) Which layer of the heart seems closest in structure to its corresponding blood vessel layer? _______________________

(c) From the descriptions given here, is the heart more similar in its wall structure to a typical artery or a typical vein? ____________

(d) Which layer of the heart wall is thickest? ____________________

(e) How does cardiac muscle differ from the smooth muscle of the blood vessels? ____

(f) Do you recall the name of the structures that connect the cardiac muscle cells? ____________________

- - - - - - - - - - - - - - - - - -

(a) tunica intima, tunica media, tunica adventitia
(b) endocardium (a layer of endothelium)
(c) artery (actually a medium artery)
(d) myocardium
(e) cardiac muscle is striated and branching
(f) intercalated discs (Chapter 2)

25. The heart directs the blood in two circulatory tracts, as shown in the diagram on the following page. The pulmonary circulation takes blood to the lungs and, after gas exchange, back to the heart. The systemic circulation takes oxygenated blood out to all the cells of the body, then eventually back to the heart. In both circulations, arteries carry blood away from the heart while veins return it.

(a) In the pulmonary circulation, blood leaves the heart from which chamber? ____________________ Into which vessel? __________ ____________________

(b) Blood from the pulmonary circulation returns to the heart through which vessel? ____________________ Into which chamber? ____________________

(c) Blood enters the systemic circulation from which chamber? ____________________ Into which vessel? ____________________

(d) Blood returns to the heart from its systemic circulation into which chamber? ____________________ Through which vessels? ____________________

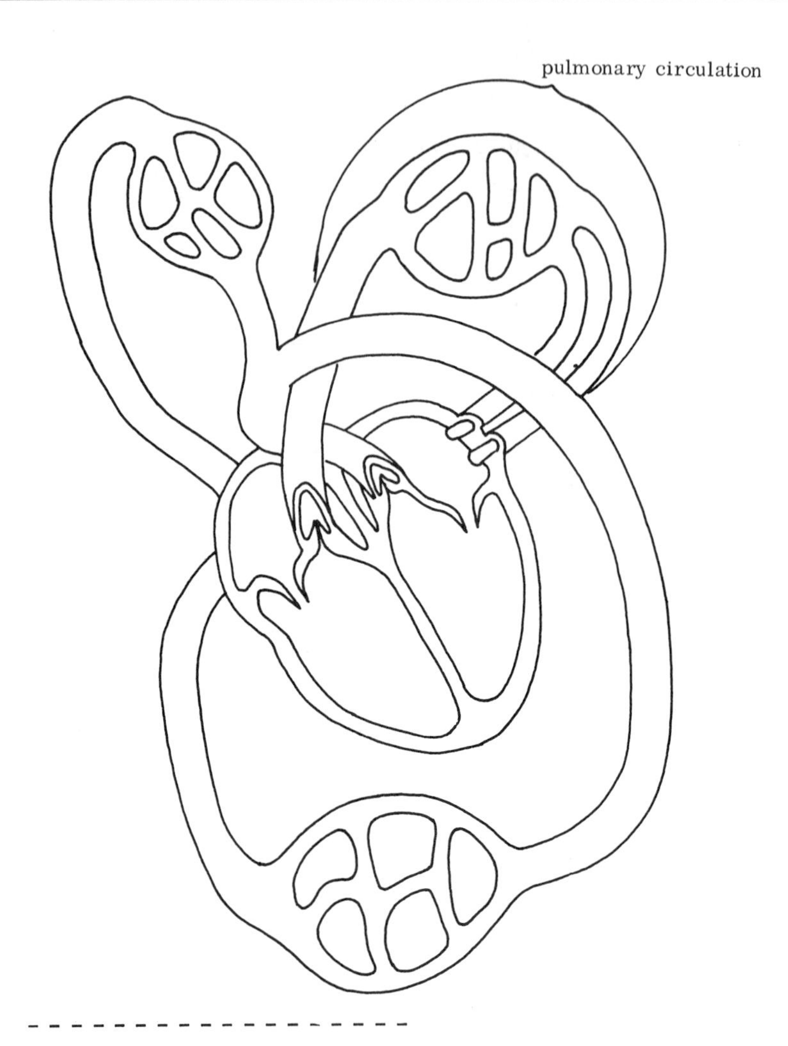

- - - - - - - - - - - - - - - - - - -

(a) right ventricle; pulmonary artery
(b) pulmonary vein; left atrium
(c) left ventricle; aorta
(d) right atrium; superior and inferior venae cavae

25. In the pulmonary circulation, the blood that leaves the heart through the pulmonary artery contains a high level of carbon dioxide. In the capillaries of the lungs, this waste gas is released and fresh oxygen is incorporated into the hemoglobin of erythrocytes. Thus oxygenated blood is returned through the pulmonary veins (there are four of them) into the left atrium.

(a) At which point in the systemic circulation would you have more highly oxygenated blood—in the left ventricle or right atrium?

_______________ In the aorta or superior vena cava?

____

(b) In the pulmonary circulation does blood contain more oxygen in arteries or veins? _______________ What about in the systemic circulation? _______________

- - - - - - - - - - - - - - - - - -

(a) left ventricle; aorta
(b) veins; arteries

The systemic circulation includes many large and important blood vessels which will be pointed out in connection with your study of the organs or tissues they supply blood to. A few of the most important are shown in the following diagram for your reference.

The brachiocephalic arteries and veins carry blood to and from the head and arms. Celiac trunk—hepatic artery and vein—supply the digestive tract. The iliac artery supplies the legs.

Several blood vessels may already be familiar to you because of their common usage. The jugular vein is in the anterior neck. Rectal veins are

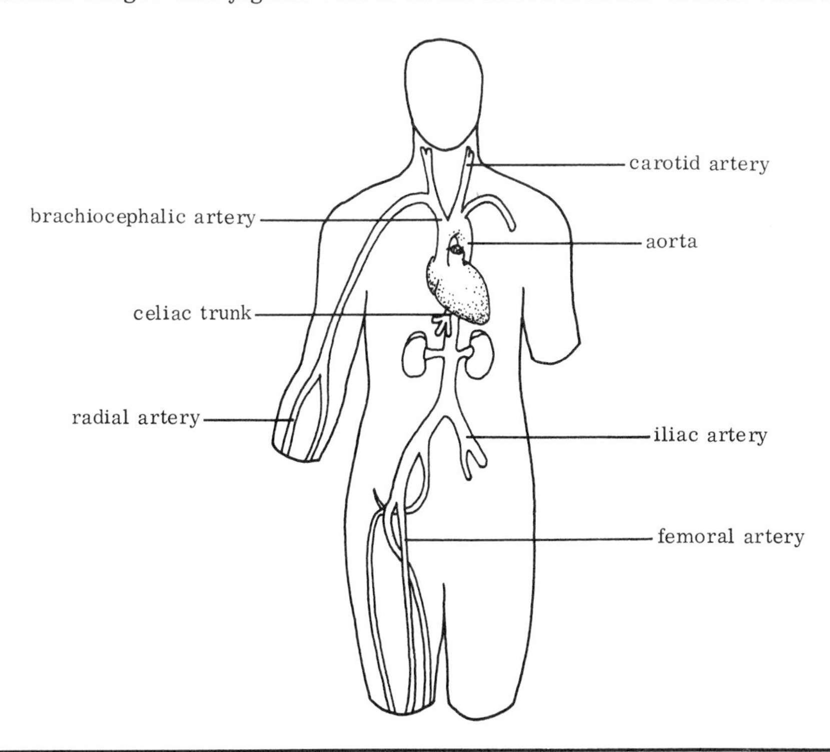

often a problem because this is where hemorrhoids occur. Saphenous veins are the sites of varicose veins of the leg. The radial artery is the spot where the pulse is taken. The brachial artery is where the blood pressure is taken.

Pulse

The radial artery pulsations reflect the beating of the heart. Thus, the pulse count is an indirect count of the number of times a heart beats per minute. A normal pulse is around 70 or 80 beats per minute.

Blood pressure

The blood pressure is a measure taken in the brachial artery of the arm. The pressure readings reflect the pressure of blood inside the heart. Two readings are taken, one measures the pressure at the peak of contraction, the other during the relaxation interval. The pressure during contraction is called systolic and may normally be around 120. The pressure during relaxation is called diastolic, and is normally around 80. A normal reading might be given as 120/80.

SELF-TEST

This Self-Test is designed to show how well you have mastered this chapter's objectives. Answer each question to the best of your ability. Correct answers and review instructions are given at the end of the test.

1. Name the three components of the circulatory system. ______________

   ____

2. Which formed element of the blood contributes to clotting? __________

3. How would you identify the element of blood that contains hemoglobin?

   ____

4. A blood vessel wall contains a thick adventitia, a few muscle fibers in its tunica media. Name the type. ____________________

5. Describe the wall of a capillary. ______________________________

6. Which blood vessel opens at the left ventricle? ____________________

7. In which circulation is oxygenated blood found in veins? ____________

8. Through which chambers does blood in the pulmonary circulation flow?

   ____

9. Name the structure indicated by A.

   ____

10. Name the structure indicated by B.

    ____

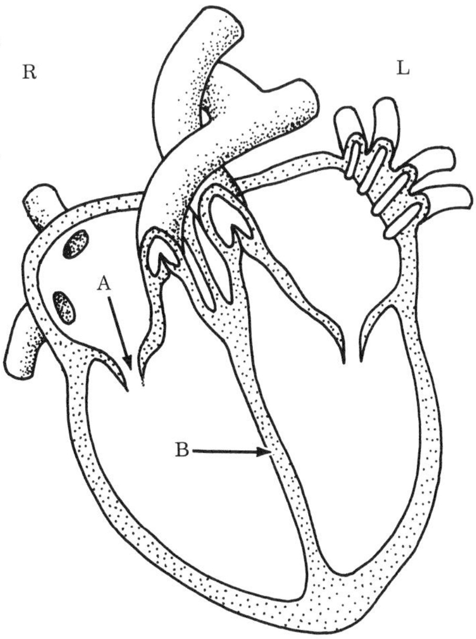

Answers

Compare your answers to the Self-Test questions with those answers given below. If all of your answers are correct, you are ready to go on to the next chapter. If you missed any, review the frames indicated in parentheses following the answers. If you missed several questions, you should probably reread the entire chapter carefully.

1. blood, blood vessels, and heart (frame 1)
2. platelet (frame 10)
3. biconcave disc, pinkish (frame 4)
4. vein (frame 18)
5. single layer of thin endothelium (frame 15)
6. aorta (frame 23)
7. pulmonary (frame 26)
8. left atrium and right ventricle (frame 25)
9. right atrioventricular valve (frame 21)
10. interventricular septum (frame 21)

CHAPTER NINE
The Lymphatic System

The lymphatic system is closely related to the circulatory system, and is sometimes considered to be a part of it. No pump moves the lymph to the body; rather, differences in pressure, along with muscles in the vessel walls, cause lymph to flow. After you complete your study of this chapter, you will be able to:

- describe the composition and derivation of lymph;
- compare lymph capillaries to blood capillaries;
- contrast lymph vessels to veins;
- compare movement of lymph to movement of blood;
- name the two major lymphatic ducts, and indicate where they join the blood circulatory system;
- identify the locations and structure of major lymph node aggregates;
- specify the contributions of the thymus, spleen, and tonsils to the lymphatic system.

LYMPH

The lymphatic system functions in maintaining the fluid balance of the body, in absorption, and in transport of substances, especially leukocytes. In this section we will examine the major component of the system, lymph, and its vessels, and consider where lymph comes from and how it gets into the vessels.

1. Lymph is generally a clear fluid, similar to plasma, containing much water and proteins; lymph also contains many leukocytes. While some lymph is manufactured in the liver, most is derived from blood. Both blood and lymph capillaries are permeable; that is, certain substances can pass through their walls. Plasma and leukocytes seep out of blood

capillaries to become tissue fluid. These substances then seep into lymph capillaries, where they are called lymph.

Erythrocytes and some large protein molecules involved in clotting cannot enter the lymphatic capillary walls. Once the fluid enters these lymph capillaries it is part of the lymphatic system.

(a) Which formed element of blood is not found in lymph? ____________

____

(b) Is lymph directly derived from blood or from tissue fluid? ________

____

(c) State one way in which the permeability of lymph capillaries differs from that of blood capillaries. ____________________

(d) Would you expect lymph to clot? ____________________

(e) Circle the substances in the following list that you would expect to find in lymph—water, small protein molecules, large protein molecules, leukocytes, erythrocytes.

- - - - - - - - - - - - - - - - - -

(a) erythrocytes
(b) tissue fluid
(c) fluid seeps out of blood capillaries but into lymph capillaries
(d) Not very well. Some large protein molecules are not in lymph. It does clot, but at a much slower rate than blood.
(e) water, small protein molecules, leukocytes

2. Lymph capillaries begin as blind-ended endothelial channels in the areas of blood capillaries. They are a little larger and a little more irregular than are the blood capillaries but the wall structure is similar.

The fluid inside the channels and that in the surrounding tissues contain different concentrations of various substances. This creates a pressure difference and causes fluid to enter the lymphatic system. In the digestive system, specialized lymphatic capillaries aid in absorption of fat, which is also added to lymph.

(a) How do the ends of lymph capillaries differ from the ends of blood capillaries? ____________________

____

(b) What causes lymph to enter the lymph capillaries? ____________

____

(c) What brings blood into blood capillaries? ____________

(d) What makes up the wall of a lymph capillary? ____________________

- - - - - - - - - - - - - - - - - -

(a) lymph capillaries have a blind end, while blood capillaries are connected to other vessels at both ends
(b) differences in pressure or osmosis
(c) arterioles, pumped by heart
(d) endothelial tissue
(The lymph channels in the digestive system will be considered in more detail in Chapter 11.)

3. Lymph capillaries join into larger and larger vessels, as do blood capillaries. The lymph vessels resemble veins, except their walls are thinner and have more valves to prevent backflow of fluid. The three layers of the walls of lymph vessels are rather indistinct, but there is definitely more muscle in the tunica media of lymph vessels than in veins.

   (a) Which would have a thicker adventitia—a vein or a lymph vessel?

   ____

   (b) Which would have more valves—a vein, an artery, or a lymph vessel? ____________________

   (c) Which would have three layers—an artery, a vein, or a lymph vessel? ____________________

- - - - - - - - - - - - - - - - - -

(a) vein; (b) lymph vessel (arteries have none); (c) all of them

Lymphatic ducts

4. All lymph vessels eventually converge into two terminal lymph ducts, the right lymphatic duct and the thoracic duct. These ducts dump lymph into the venous blood at the right and left subclavian veins. The right lymphatic duct collects lymph from the right arm, right side of the head and chest, and the right lobe of the liver. All other areas send their lymph to the thoracic duct. These terminal lymphatic ducts are similar in structure to the other lymphatic vessels, but have thicker walls.

   (a) Which lymphatic duct dumps lymph into the left subclavian vein?

   ____

(b) Through which lymphatic duct would lymph from the right axilla be added to the blood? ____________________

(c) Through which lymphatic duct would lymph from the right leg be added to the blood? ____________________

(d) How many layers make up the walls of the lymphatic ducts? ______

- - - - - - - - - - - - - - - - - - -

(a) thoracic duct; (b) right lymphatic duct; (c) thoracic duct; (d) three

5. What causes lymph to flow from the capillaries to where it rejoins the blood circulatory system? As we saw in the discussion of capillaries, pressure is the major factor. Pressure is effective as a mover of lymph especially in the capillaries and in the ducts. (In the next chapter we will deal more with pressures in the chest.) The massaging of lymph vessels by contractions of skeletal muscles and arterial walls also helps move lymph forward, as does the continuous formation of new lymph in the capillaries. The smooth muscle in walls of lymphatic vessels contracts also, forcing lymph to move forward as the valves prevent its backflow.

(a) What effect does cardiac muscle have on lymph flow? __________

____

(b) Do skeletal and smooth muscle have any effect on lymph flow?_____

____

(c) What is the primary factor in lymph flow? ____________________

(d) What prevents lymph from flowing from lymph vessels into lymph capillaries? ____________________

(e) If lymph is being formed very rapidly, why wouldn't it just seep out of the lymph capillaries and rejoin the tissue fluid? __________

____

- - - - - - - - - - - - - - - - - - -

(a) no direct effect (or very little)
(b) yes, they massage lymph through vessels
(c) pressure
(d) valves
(e) lymph capillaries are only permeable in one direction because of the pressure difference

6. Answer the following questions for a summary of lymph, vessels, and flow.

   (a) List the three major constituents of lymph. ____________________

   ____

   (b) Would water and leukocytes be more likely to pass out the walls of a blood capillary or a lymph capillary? ____________________

   (c) In its function, is a lymphatic vessel more similar to an artery or a vein? ____________________

   (d) Into which veins do the lymphatic ducts empty lymph? ____________

   ____

   (e) How does the wall of a large lymphatic vessel differ from that of a large vein? ____________________

   (f) How does a lymphatic capillary differ from a blood capillary in its terminations? ____________________

   In its permeability? ____________________

- - - - - - - - - - - - - - - - - -

(a) water, small protein molecules, and leukocytes
(b) blood capillary
(c) vein (returns fluid to heart)
(d) right and left subclavian veins
(e) lymphatic tissue wall is generally thinner, but has more muscle
(f) lymphatic capillary has one blind end; fluids only diffuse in

Lymph nodes

7. Along the course of lymphatic vessels, small oval structures called lymph nodes are found. These act as filters for the lymph, serving to purify it before it is returned to the bloodstream. Several afferent lymphatic vessels enter each lymph node, while only one or two efferent vessels leave it. The diagram on the following page shows the major groups of lymph nodes.

   (a) Would you expect more lymphatic vessels to enter or leave an axillary lymph node? ____________________

   (b) Is a lymphatic vessel that carries lymph into a cervical lymph node considered afferent or efferent? ____________________

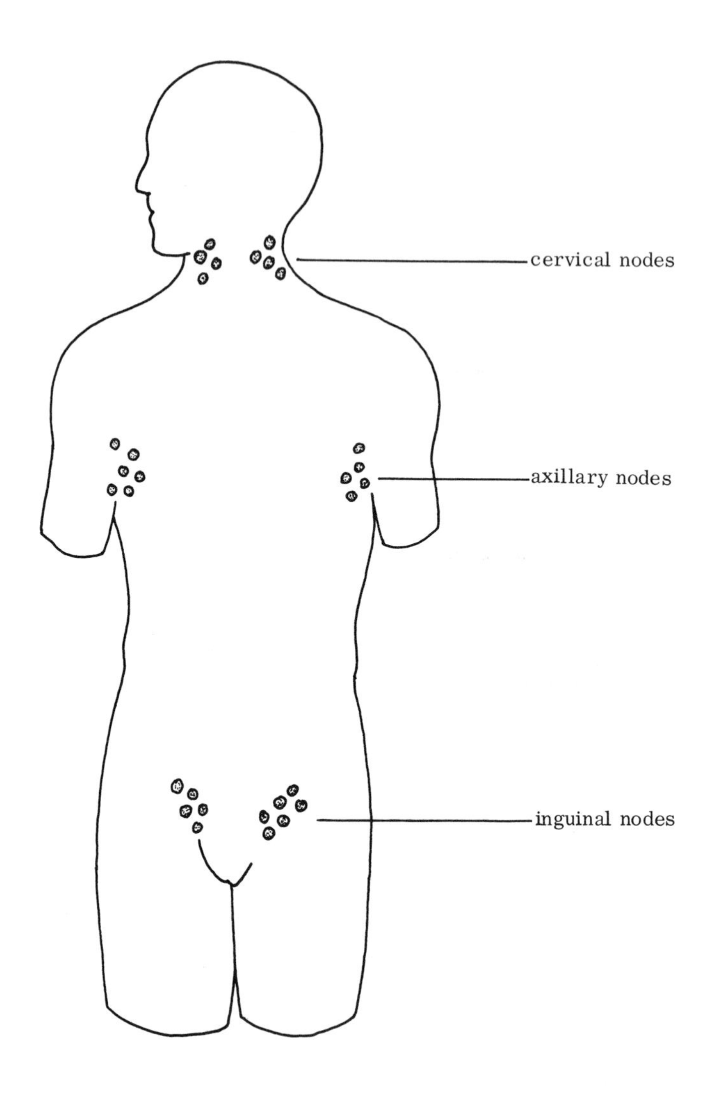

(c) What is the primary function of a lymph node? ____________________

(d) Is a lymph node shaped more like an erythrocyte, an egg, or a cube?

____

- - - - - - - - - - - - - - - - - - - -

(a) more enter, only one or two leave any node
(b) afferent (approach = afferent; exit = efferent)
(c) filter
(d) egg

8. The major groups of lymph nodes are located bilaterally in the neck region, the armpit, and groin.

   (a) What are these groups called? ______________________________

   (b) Which of the major groups of lymph nodes empty into the thoracic duct? ______________________________

- - - - - - - - - - - - - - - - - -

(a) cervical, axillary, and inguinal
(b) left cervical, left axillary, left inguinal, and right inguinal

9. Each lymph node is made up of "lymphatic tissue" surrounded by a fibrous connective tissue capsule. As lymph enters through the afferent lymphatics, it filters through the lymphatic tissue. Any elements that may have entered the lymph due to disease or trauma, such as erythrocytes and bacteria, are removed, while the lymphatic tissue adds lymphocytes and immunity factors to the outgoing lymph. The products of cancers often become enmeshed in the tissues of lymph nodes, which facilitates their growth and spread in the body.

   (a) Would antibodies be filtered from the blood or added to it in lymph nodes? ____________________

   (b) Are lymph vessels continuous through a lymph node? ____________
   ____

   (c) What is the effect of a lymph node on bacteria? ________________

- - - - - - - - - - - - - - - - - -

(a) added to it
(b) no, lymph is filtered through lymphatic tissue before leaving the node
(c) filters it out (phagocytosis)

In performing their filtration function, lymph nodes often become inflamed as a result of bacterial buildup. This is especially evident in the cervical nodes, which are often referred to as "swollen glands" when the neck nodes can be felt. Bits of cancerous tissue also may build up in lymph nodes. The axillary nodes may receive cancerous tissue from breasts, while the inguinal nodes reflect the state of pelvic cancer. When a cancer has spread to the lymph nodes, a chance for cure is reduced.

LYMPHOID ORGANS

The lymphatic system proper consists of lymph, lymph vessels, and lymph nodes. Several other organs, however, are closely associated with the lymphatic system. In this section, we will discuss the structure and function of lymphoid organs, the spleen, the thymus, and the tonsils.

10. The spleen is located in the left upper abdomen, beneath the diaphragm and behind the stomach. It is a soft, ovoid structure, about 5 inches by 3 inches when at rest. The spleen is covered with an elastic, muscular capsule having a sponge-like appearance, with many sinusoids, or widened channels, for blood. The body of it is divided into compartments.

    (a) Indicate the location of the spleen in the figure.

    (b) How does the splenic capsule differ from a lymph node capsule? ____________________

    ____

    ____

    ____

    (c) What is the effect of blood sinusoids on the texture of the spleen? ____________________

    ____

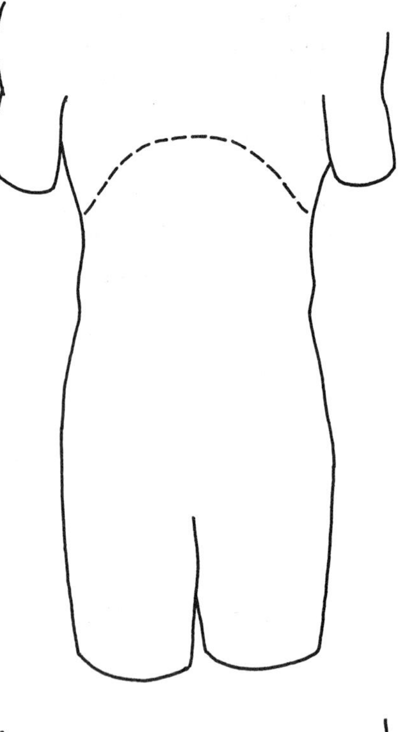

- - - - - - - - - - - - - - - - - -

(a) (See figure.)
(b) splenic capsule is elastic and muscular; lymph node capsule is fibrous connective tissue
(c) make it soft and spongy

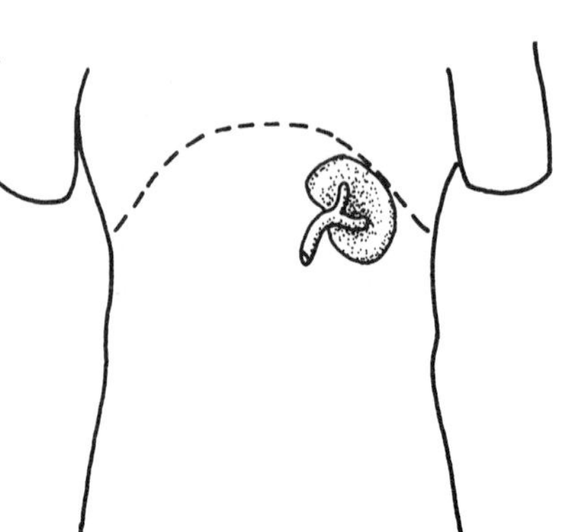

11. The walls of the sinusoids of the spleen are made up of widely spaced phagocytic cells that ingest material to be removed from blood. Blood enters these sinusoids through the splenic artery from the aorta. Much of it escapes through the sinusoid walls to be processed in the lympoid tissue of the spleen. The spleen thus functions in filtering the blood, much as the lymph nodes filter the lymph. Aged erythrocytes are destroyed in the spleen, and a great deal of blood can be stored here, due to its elastic nature. As elsewhere in the lymphatic system, factors critical to the body's immune defense system are added to blood by the spleen.

    (a) What might be the function of the phagocytic cells in sinusoid walls in the spleen? ____________________

    (b) What vessel would you expect to carry blood away from the spleen?

    ____

    (c) Complete the following three functions of the spleen.

    Filter ____________________

    Destroy ____________________

    Add ____________________

- - - - - - - - - - - - - - - - - -

(a) digest bacteria
(b) splenic vein
(c) filter blood; destroy erythrocytes; add immunity factors to blood

12. A second lymphoid organ, the thymus, is located high in the chest between the sternum and the aorta, just below the thyroid gland. This organ is relatively large at birth and continues to grow until adolescence. Children who have deficient thymic tissue are very susceptible to infection and allergy. The thymus is a prime producer of lymphocytes with special immunologic properties. These thymus-produced lymphocytes seem to survive longer than others. The thymus has only efferent lymphatic vessels—it serves no filtration function.

    (a) What is the primary function of the thymus? ____________________

    ____

(b) Indicate the location of the thymus in the figure.

(c) Lymph nodes filter lymph and the spleen filters blood. What does the thymus filter?

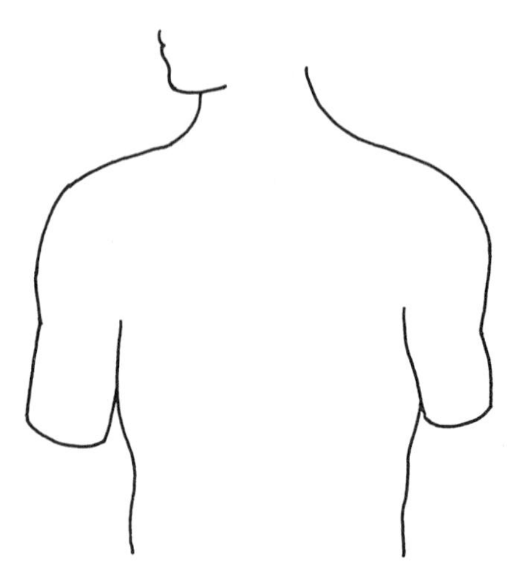

- - - - - - - - - - - - - - - - - -

(a) produces lymphocytes to aid in immunity
(b) high in the chest, below thyroid (See figure at right.)
(c) nothing

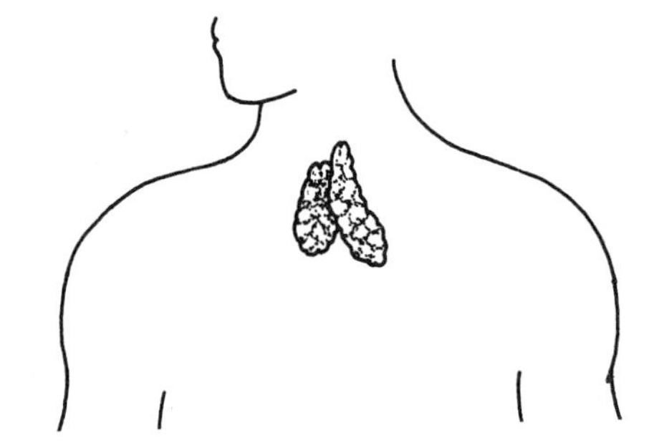

13. The tonsils are lymphoid organs in the oral and nasal cavities that contribute to lymphocyte production. The two largest tonsils are the palatine tonsils, one on each side at the back of the throat. These are the tonsils removed in a typical tonsillectomy. In the nasal cavity, above the soft palate, is the pharyngeal tonsil, typically removed in an operation called an "adenoidectomy." A third type of tonsillar tissue is incorporated into the base of the tongue; this is called the lingual tonsil and is seldom, if ever, removed. These tonsils form a protective ring around the entrance to the respiratory and digestive tracts, their phagocytic lymphocytes ready to attack bacteria.

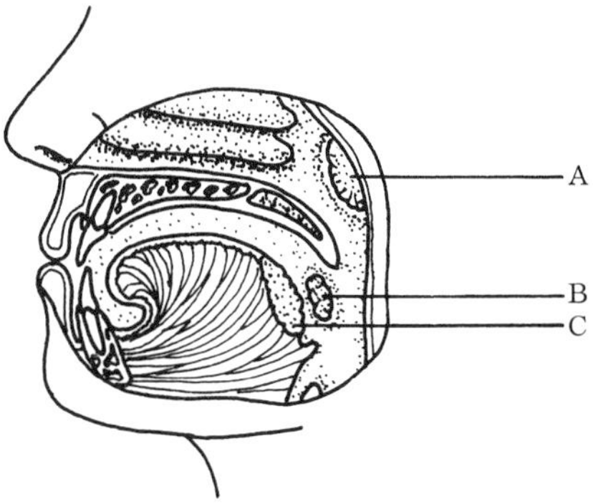

(a) Name the three tonsils indicated in the drawing.

A ______________________

B ______________________

C ______________________

(b) Which tonsils are removed in a "tonsillectomy and adenoidectomy," or a standard T&A?

____

____

(c) What is manufactured in the tonsils? ______________________

- - - - - - - - - - - - - - - - - -

(a) A—pharyngeal; B—palatine; C—lingual
(b) pharyngeal and palatine
(c) lymphocytes

14. Answer the following questions for a review of lymphoid organs.

(a) What lymphoid structures are located in the groin? ______________

____

(b) What lymphoid organ destroys erythrocytes? ______________________

(c) What function do the thymus and tonsils have in common? ________

____

(d) Which lymphoid structures act as a filter for lymph? ____________

____

- - - - - - - - - - - - - - - - - -

(a) inguinal lymph nodes; (b) spleen; (c) produce lymphocytes;
(d) lymph nodes

SELF-TEST

This Self-Test is designed to show how well you have mastered this chapter's objectives. Answer each question to the best of your ability. Correct answers and review instructions are given at the end of the test.

1. List the three major components of lymph. ______________________

____

2. From where is most of the lymph derived? ______________________

____

3. Name two factors that cause lymph to flow. ______________________

____

4. Name the duct that returns lymph from the left arm to the blood. _____

____

5. A structure with a fibrous connective tissue capsule, several afferent lymphatics, and one efferent lymphatic vessel would be ____________

____

6. Name and locate three major groups of lymph nodes. ______________

____

7. Which lymphoid organ filters the blood? ______________________

8. What is the basic function of the thymus? ______________________

____

9. Identify the structure A in the diagram. ________________

10. What blood vessel takes blood away from structure B in the diagram? ________________

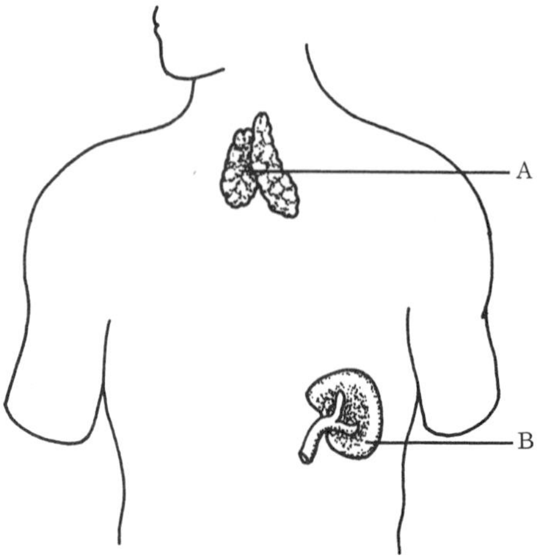

A

B

Answers

Compare your answers to the Self-Test questions with those answers given below. If all of your answers are correct, you are ready to go on to the next chapter. If you missed any, review the frames indicated in parentheses following the answers. If you missed several questions, you should probably reread the entire chapter carefully.

1. water, small protein molecules, and leukocytes (frame 1)
2. indirectly from blood; and directly from tissue fluid (frame 1)
3. pressure, and action of adjacent muscle (frame 5)
4. thoracic duct (frame 4)
5. a lymph node (frame 9)
6. cervical, in the neck; axillary, in the armpit; inguinal, in the groin (frame 8)
7. spleen (frame 10)
8. produce lymphocytes with immunologic factors (frame 12)
9. thymus (frame 12)
10. splenic vein (frame 11)

CHAPTER TEN
The Respiratory System

The body needs a continual ventilation system to maintain its supply of oxygen on a cellular level. This system carries on the process known as external respiration—the exchange of oxygen and carbon dioxide between the blood and the outside air. Internal respiration refers to the exchange of the gases between blood and body tissues. After you complete your study of this chapter, you will be able to:

- identify and locate the organs of the respiratory system;
- specify the characteristics of respiratory epithelium;
- differentiate between the trachea and bronchi;
- compare the pressures in different parts of the thoracic cavity;
- identify the major nerves and blood vessels that supply the lungs;
- specify the levels of conducting tubes that lead air to respiratory bronchioles;
- trace the path of an oxygen molecule from the air to a body cell;
- list at least three protective mechanisms in the respiratory system.

The organs of respiration include the nose, the pharynx, the larynx, the trachea, the bronchi (singular, bronchus), and the lungs, as shown on the next page. These are divided into the conducting system, in which air is transported, and the respiratory system, in which gases are exchanged. In the first section, we will examine the conducting portion of the respiratory system.

THE CONDUCTING PORTION

1. The nose begins the respiratory tract, as it is in direct contact with the outside air. Here the air is filtered through nasal hairs, warmed by superficial blood vessels, and moistened by secretions of mucous cells. Here also are the olfactory receptor cells we examined in Chapter 6.

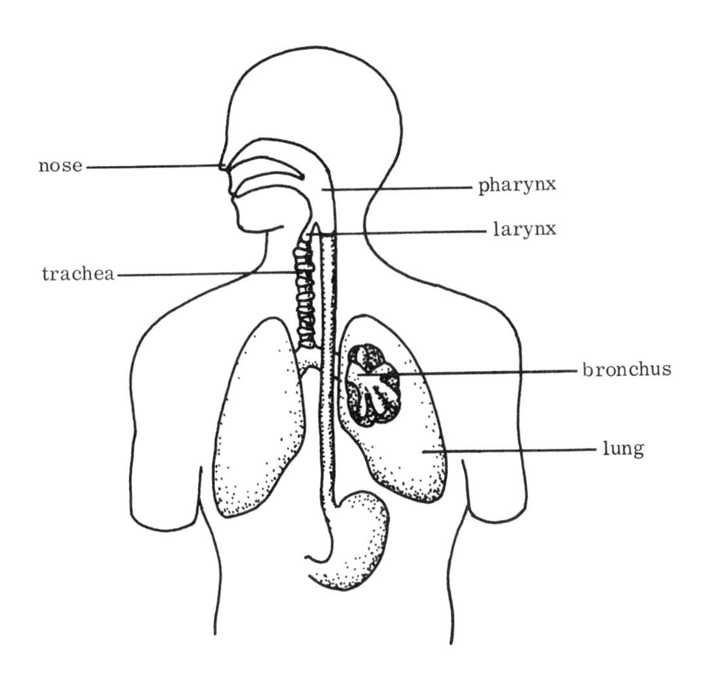

Most of the interior of the nose is covered by respiratory epithelium. The tissue is pseudostratified columnar in type. The cells contain cilia on their free ends, which beat rhythmically to sweep microscopic particles back out of the respiratory tract. Goblet-shaped cells secrete mucous, which aids in trapping particles for removal.

(a) In the internal nose, what moistens the air? ____________________

(b) What removes large particles from inhaled or inspired air? ______

____

(c) What removes small particles from inspired air? ______________

(d) Give two functions of the mucous secreted by goblet cells. _______

____

(e) What contributes to warming air? ___________________________

- - - - - - - - - - - - - - - - - -

(a) mucous; (b) hairs in the nose; (c) cilia; (d) trap small particles and moisten air; (e) blood vessels

2. Internally, the nose is divided into two nasal cavities by the nasal septum. The septum is mainly cartilage, as is most of the structure of the external nose. The outer sides of the nasal cavities, however, contain

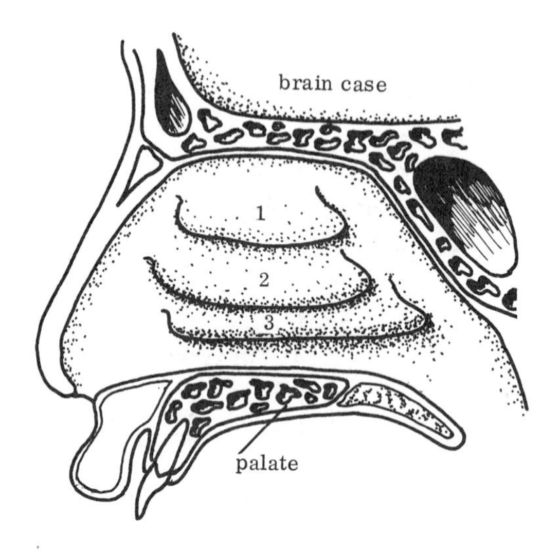

small bones underlying the respiratory epithelium. These bones create an uneven surface, which causes air turbulence; this, in turn, exposes more of the inspired air to the beneficial effects of warming, moistening, and filtering.

The bony projections, as numbered in the diagram, are (1) superior nasal concha, (2) middle nasal concha, and (3) inferior nasal concha. As you can see from the diagram above, the nasal cavities are bounded superiorly by the brain case and inferiorly by the palate.

(a) Consider the right nasal cavity. Would the lateral or medial wall be cartilaginous? ____________________

(b) The nasal septum is bounded superiorly by the brain case. What is it bounded by inferiorly? ____________________

(c) Near which bony projection in the nose would you expect to find olfactory receptors? ____________________

(d) What is the effect of the nasal conchae on inspired air? __________ ____________________

- - - - - - - - - - - - - - - - - -

(a) medial (left); (b) palate; (c) superior nasal concha; (d) create air turbulence

3. The posterior nasal cavities are continuous with the superior portion of the pharynx, the nasal pharynx. The nasal pharynx extends from the

back wall of the nasal cavities to the border of the soft palate. The middle portion of the pharynx, the oral pharynx, extends from the border of the soft palate to the hyoid bone. The laryngeal pharynx extends from the level of the hyoid bone to the point where the digestive tract begins. Both the oral and laryngeal portions of the pharynx serve the digestive system as well as the respiratory system, while the nasal pharynx is solely devoted to ventilation.

(a) Name the divisions of the pharynx indicated in the drawing at the right.

A ____________________

B ____________________

C ____________________

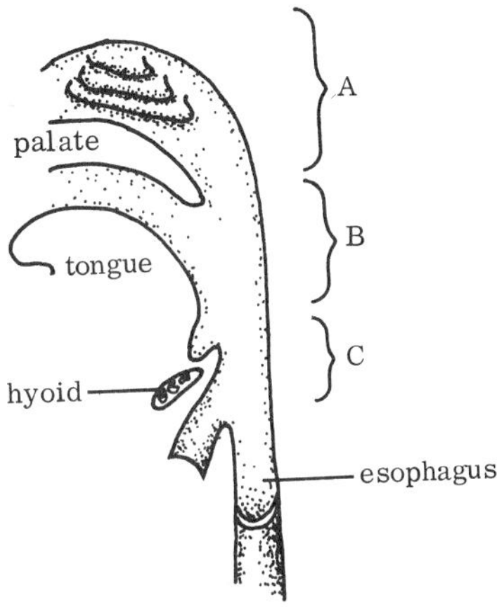

(b) Which division of the pharynx contains the palatine tonsils?

____

(c) Which division of the pharynx contains the pharyngeal tonsil?

____

- - - - - - - - - - - - - - - - - -

(a) A—nasal pharynx; B—oral pharynx; C—laryngeal pharynx
(b) oral pharynx
(c) nasal pharynx

4. The pharynx serves a dual function in conducting both air and food. The epithelial lining reflects this as the nasal pharynx has pseudostratified epithelium with cilia. The oral and most of the laryngeal pharynx are covered with a different type, stratified squamous epithelium. Persons who breathe through their mouths do not benefit from the "air conditioning" process provided by the nose and nasal pharynx.

(a) What type of epithelium would be found in the portion of the pharynx that extends from the hyoid bone to the level of the opening of the esophagus? ____________________

(b) What portion(s) of the pharynx contain(s) respiratory epithelium?

____

(c) What is the inferior border of the portion of the pharynx that does not contribute to the digestive system? ____________________

(d) Name three processes that occur in the nasal cavity that are not possible when air enters the respiratory system through the mouth.

____

- - - - - - - - - - - - - - - - - -

(a) stratified squamous; (b) nasal pharynx; (c) soft palate; (d) warming, moistening, and filtering air

5. At the inferior border of the laryngeal pharynx, the esophagus begins the digestive system proper. Here also is found the larynx. The larynx, or voice box, is made up of several cartilages. One of these, the epiglottis, covers the opening into the larynx before swallowing. The epiglottis thus keeps the respiratory system free from any solid or liquid matter that might interfere with breathing. The thyroid cartilage makes up the front and side walls of the larynx and is commonly known as the Adam's apple. Other cartilages with laryngeal muscle and epithelial coverings make folds within the larynx. Vibrations of these folds, or vocal cords, produce the sounds we know as speech.

   (a) Most of the larynx itself is lined with respiratory epithelium. What type is it? ______________________________

   (b) Name the large cartilage that makes up the walls of the larynx.

   ____

   (c) What is the function of the epiglottis? ______________

   ____

   (d) What do you think would cause vocal cords to vibrate? ________

   ____

- - - - - - - - - - - - - - - - - -

(a) pseudostratified columnar with cilia and goblet cells; (b) thyroid cartilage; (c) cover entrance to respiratory tract during swallowing; (d) passage of air

6. Directly inferior to the larynx is the beginning of the trachea. The trachea, or windpipe, extends straight down the neck and center of the chest for about 11 cm. The tube is almost an inch in diameter and the shape is maintained by 16 to 20 C-shaped cartilaginous rings, completed in the back by the trachealis muscle. The lining is typical respiratory epithelium.

   The trachea divides into two primary bronchi, one for each lung.

The right bronchus is more nearly vertical than the left, and only half as long. The lining and cartilage are similar in bronchi and trachea.

(a) Name the structures indicated in the diagram at the right.

A ____________

B ____________

C ____________

(b) If you inhaled a piece of popcorn, in which bronchus would it more likely lodge? ______

____

(c) What keeps the trachea open for continual passage of air?

____

____

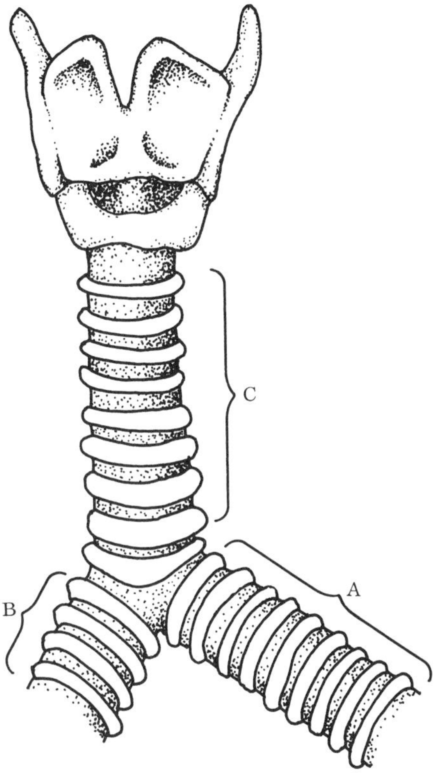

- - - - - - - - - - - - - - - - - - -

(a) A—left bronchus; B—right bronchus; C—trachea
(b) right (more nearly vertical)
(c) C-shaped cartilages

7. The bronchi begin to subdivide into smaller and smaller tubules. These are then called secondary and tertiary bronchi, bronchioles, and terminal bronchioles. As the tubes get smaller, the cartilage changes from rings to plates, which get smaller until no cartilage is seen in bronchioles. Smooth muscle appears as cartilage begins to diminish, so more smooth muscle is present in walls of terminal bronchioles than in bronchioles or bronchi. The epithelium, too, gradually changes as goblet cells cease to appear and then pseudostratified cells are replaced by simple columnar cells in terminal bronchioles. Cilia remain constant, however.

(a) What cartilage configuration would you expect to see in the smallest bronchi? ______________

(b) Would you expect to see more smooth muscle in the wall of a secondary bronchus or a terminal bronchiole? ______________

(c) In what tubes would you expect to see pseudostratified ciliated epithelium without goblet cells? ______________

(d) What type of epithelium is present in terminal bronchioles? ______________ Are there cilia? ______________ Goblet cells? ______________

- - - - - - - - - - - - - - - - - -

(a) very small plates or chips
(b) terminal bronchiole
(c) bronchioles
(d) simple columnar; yes, there are; no goblet cells

8. Answer the following questions for a review of the conducting portion of the respiratory system.

(a) Identify the structures indicated in the figure at the right.

A ______________

B ______________

C ______________

D ______________

E ______________

F ______________

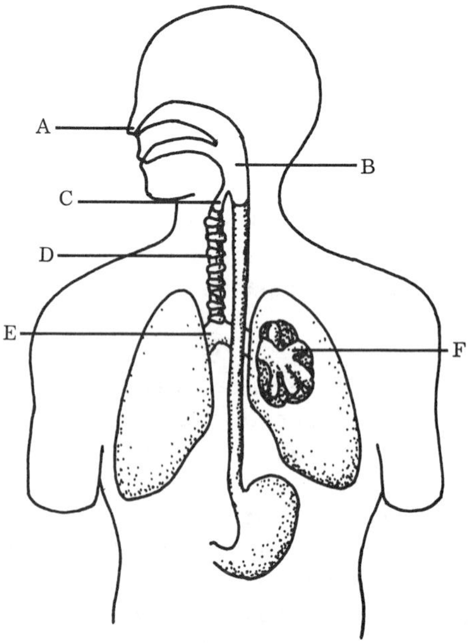

(b) What are the superior and inferior borders of the oral pharynx? ______________

____

(c) Describe typical respiratory epithelium. ______________

____

(d) Name two organs or portions of the respiratory system where the epithelium is not typical respiratory epithelium. ________________

____

(e) What is the result of turbulence in the air created by the nasal conchae? ________________________________

(f) How do the right and left bronchi differ structurally? ____________

____

(g) What two different structures form and maintain the shape of the trachea? ________________________________

(h) Specify the characteristics of a terminal bronchiole in terms of:

(1) epithelium ____________________

(2) cartilage ____________________

(3) smooth muscle ________________

- - - - - - - - - - - - - - - - - -

(a) A—nose; B—pharynx; C—larynx; D—trachea; E—right bronchus; F—secondary bronchus
(b) soft palate; hyoid bone
(c) pseudostratified (columnar) with cilia and goblet cells
(d) oral pharynx; laryngeal pharynx; terminal bronchiole (any two)
(e) more air is warmed, moistened, filtered
(f) right is more vertical, shorter
(g) C-shaped cartilage; trachealis muscle
(h) (1) ciliated columnar epithelium; (2) no cartilage; (3) much smooth muscle

THE RESPIRATORY PORTION

9. The conducting portion of the respiratory system includes up to 70 generations of branches down to the level of the terminal bronchioles. Below this level, the respiratory portion begins. Through these levels of subdivisions, the epithelium gradually thins from columnar to non-ciliated simple squamous. The muscle in the walls of the respiratory bronchioles is gradually replaced by elastic connective tissue, and outpouchings called alveoli (singular, alveolus) are seen along the walls. The diameter of the tubes in the respiratory division is relatively constant in spite of continued branching.

(a) In the wall structure of a terminal bronchiole, what type of epithelium is found? ________________ What type of cartilage, muscle, or elastic connective tissue? ________________

(b) Are outpouchings of the walls found in terminal bronchioles or respiratory bronchioles? ________________

(c) What are these outpouchings called? ________________

(d) What type of bronchiole would have elastic connective tissue lined with simple squamous epithelium? ________________

- - - - - - - - - - - - - - - - - -

(a) ciliated columnar epithelium; smooth muscle in walls
(b) respiratory
(c) alveoli
(d) respiratory

10. The alveoli of the respiratory bronchioles are often clustered around alveolar ducts. These clusters are then surrounded by capillary beds. The increased vascularity at this level is necessary for an efficient gas exchange system through the large surface area.

(a) Identify the structures shown at the right.

A ________________

B ________________

C ________________

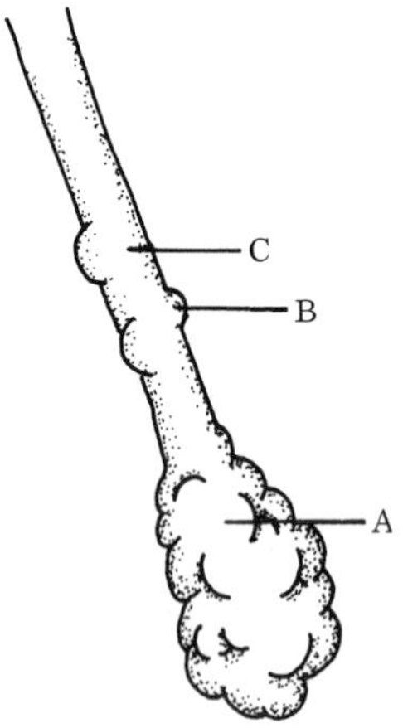

(b) Are the capillaries that surround the respiratory alveoli part of the systemic or pulmonary circulatory system?

____

(c) What type of epithelium lines the alveoli?

____

- - - - - - - - - - - - - - - - - -

(a) A—alveolar duct; B—alveoli; C—respiratory bronchiole
(b) pulmonary
(c) simple squamous

11. All the subdivisions of the bronchi are contained within the organs we call lungs. The two lungs are enclosed in the thoracic cavity and separated by the mediastinum. The trachea and bronchi are located in the mediastinum along with the heart, aorta, venae cavae, and thymus.

    Each lung is covered with a two-layered membrane, called the pleura. One layer is closely adherent to the muscular thoracic wall; this layer is called the parietal pleura. The other layer is closely adherent to the lung; it is the visceral pleura. These layers contain a fluid potential space, called the pleural cavity, that slides easily as breathing takes place. The adhesive effect of the visceral pleura also helps to keep the lung expanded. Each lung has a costal surface along the rib cage, an apex at the superior aspect, and a diaphragmatic surface at the inferior aspect.

    (a) Identify the structures shown at the right.

    A ____________________

    B ____________________

    C ____________________

    (b) Name the portion of the thoracic cavity in which the trachea and bronchi are found.

    ____

    (c) How much space exists between the visceral and parietal pleura? ____________________

    (d) Label the surfaces of the lung indicated at the right.

    A ____________________

    B ____________________

    C ____________________

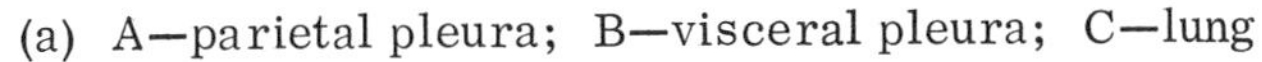

(a) A—parietal pleura; B—visceral pleura; C—lung
(b) mediastinum
(c) very little (potential space)
(d) A—costal; B—diaphragmatic; C—apex

12. The right lung is divided into three lobes, the left into two. The left lung is smaller, as the heart takes up more space on that side.

    The pulmonary arteries bring blood from the heart to the capillary beds near alveoli for oxygenation. Bronchial arteries from the circulatory system branch from the aorta and nourish the tissues of the conducting division of the respiratory system. The pulmonary capillaries nourish the respiratory division, before returning the blood to the heart via the pulmonary veins.

    (a) Which artery supplies terminal bronchioles? ____________

    (b) Which vein takes blood away from alveoli? ____________

    (c) From which chamber of the heart does the vessel lead that nourishes the respiratory bronchioles? ____________

    (d) Blood leaving the lungs via the pulmonary circulation enters which chamber of the heart? ____________

    (e) Blood leaving the lungs via the systemic circulation returns through which vein? ____________

- - - - - - - - - - - - - - - - - -

(a) bronchial artery; (b) pulmonary vein; (c) right ventricle; (d) left atrium; (e) bronchial vein

Mechanics of external respiration

13. The pleural cavity is not open to the atmosphere. This cavity normally has a lower pressure than the outside air. This causes some air to remain in the lungs.

    Inspiration occurs when certain sets of muscles contract. The intercostal muscles between the ribs contract, causing the rib cage to become larger. At the same time, the diaphragm contracts, lengthening the thoracic cavity by pulling down its floor. The parietal pleura adheres to the thoracic wall, the visceral pleura adheres to the lung itself, and the pressure difference causes the lung to expand as air rushes in. When the diaphragm and intercostal muscles relax, the air is forced out of the lungs. About a half liter of air is brought in and expired during each breath.

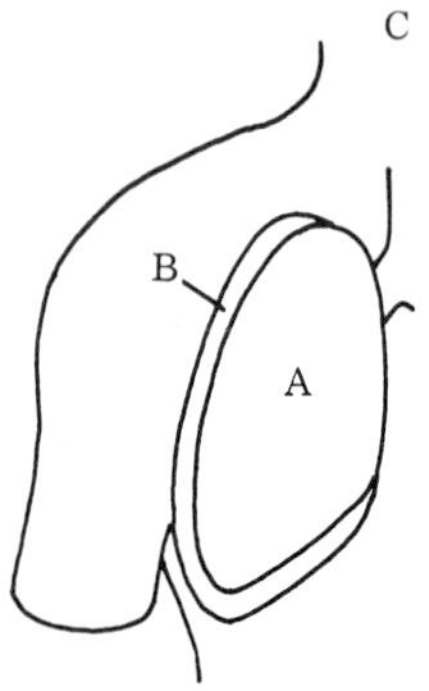

    C

    B

    A

    (a) At which point in this diagram (A, B, or C) is pressure lowest? ____________

    (b) Which two sets of muscles contract during inspiration? ____________

(c) What tissue in the lung do you think allows it to expand and contract?

____

- - - - - - - - - - - - - - - - - -

(a) B—pleural cavity; (b) diaphragm and intercostal muscles;
(c) elastic connective tissue

14. During inspiration, oxygen-laden air enters the alveoli surrounded by capillary beds. Oxygen molecules must pass through the thin squamous epithelium of the alveolar wall, through a small amount of tissue space, and through the endothelium of the adjacent capillary. Then it dissolves in the plasma or is incorporated into an erythrocyte to continue its journey to body tissues. Ninety-five percent of the gas is carried in hemoglobin. Carbon dioxide follows the same path in reverse.

(a) List in order the structures a carbon dioxide molecule passes through on its way from an erythrocyte to the air in the trachea. We have filled in some of the structures.

(1) plasma ________ (2) ________________

(3) ________________ (4) ________________

(5) respiratory bronchiole ________ (6) ________________

(7) ________________ (8) ________________ (9) trachea

(b) Is an oxygen molecule more likely to be transported dissolved in plasma or as part of an erythrocyte? ________________

____

- - - - - - - - - - - - - - - - - - -

(a) (2) capillary endothelium, (3) tissue space, (4) alveolar wall, (6) terminal bronchiole, (7) bronchiole, (8) bronchus
(b) combined with hemoglobin as part of an erythrocyte

15. Respiration is a reflex action, but is under some voluntary control. The major nerves controlling the process are the intercostal, the phrenic (controls diaphragm) and the vagus nerve. The centers of respiration are in the medulla oblongata of the brain.

(a) In which part of the brain is respiratory control localized? ________

____

(b) Which nerve stimulates the intercostal muscles? ____________

The diaphragm? ____________

- - - - - - - - - - - - - - - - - - -

(a) medulla oblongata
(b) intercostal; phrenic

16. Reflexive mechanisms also help protect the respiratory system from foreign invasion. A sneeze takes place when the nasal lining is irritated. A cough results from an irritation of the tracheal or bronchial lining. Both have a forced expiration that attempts to remove the source of irritation from the system. These reflexes, with the nasal cavity conditioning of air, serve to protect the respiratory system from harm.

(a) What conditioning of air takes place in the nasal cavity? ____________

____

(b) If a particle escapes the hair and cilia filters of the upper respiratory tract, and irritates the lining of the right bronchus, what would occur?

____

(c) What is the effect of a sneeze? ____________

- - - - - - - - - - - - - - - - - - -

(a) warm; moisten; filter
(b) cough
(c) force particles out of nose

SELF-TEST

This Self-Test is designed to show how well you have mastered this chapter's objectives. Answer each question to the best of your ability. Correct answers and review instructions are given at the end of the test.

1. What tissues make up the wall of a respiratory bronchiole? ____________

   ____

2. In which part of the thoracic cavity is pressure normally lowest? ______

   ____

3. What blood vessel supplies the secondary bronchioles? ________________

   ____

4. What blood vessel entering or leaving the lung has the highest oxygen content? ________________________

5. Identify the structures shown at the right.

   A ________________________

   B ________________________

   C ________________________

   D ________________________

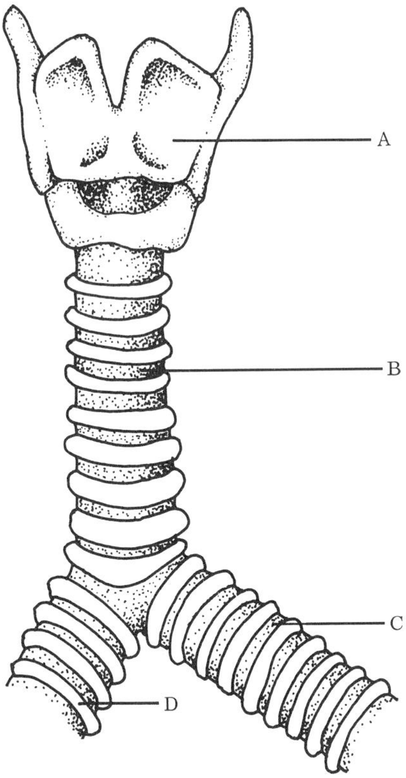

6. Arrange the structures listed below in the sequence an oxygen molecule would pass through on its way from the air to the left ventricles of the heart. We have filled in the first one of the sequence.

   ___1___ nasal pharynx

   ______ right ventricle

   ______ capillary endothelium

   ______ larynx

   ______ bronchus

   ______ alveolar wall

   ______ plasma

   ______ erythrocyte

7. Which two nerves directly cause an inspiration? ______

   ______

8. Where in the respiratory system would you find epithelium of the following types:

   (a) pseudostratified columnar with cilia and goblet cells overlying conchae ______

   (b) columnar ciliated with goblet cells ______

9. What are the superior and inferior limits of the oral pharynx? ______

   ______

10. Name three protective mechanisms found in the respiratory system.

    ______

    ______

Answers

Compare your answers to the Self-Test questions with those answers given below. If all of your answers are correct, you are ready to go on to the next chapter. If you missed any, review the frames indicated in parentheses following the answers. If you missed several questions, you should probably reread the entire chapter carefully.

1. simple squamous epithelium and elastic connective tissue (frame 9)
2. pleural cavity (frame 13)
3. bronchial artery (frame 12)
4. pulmonary vein (frame 12)
5. A—larynx (thyroid cartilage); B—trachea; C—left bronchus; D—right bronchus (frames 5 and 6)
6. 1, 8, 5, 2, 3, 4, 6, 7 (frames 7 and 14)
7. phrenic and intercostal (frame 15)
8. (a) nasal cavity (frame 2)
   (b) terminal bronchiole (frame 7)
9. soft palate and hyoid bone (frame 4)
10. filtering cilia, mucous for moistening, reflex for removing particles by sneeze or cough, epiglottis (any three) (frames 1, 5, and 15)

CHAPTER ELEVEN
The Digestive System

The digestive system performs a critical maintenance function for the body. The task of converting the food we eat into nutrients to nourish all the cells of the body is reserved for the digestive system. We shall see how this system works with the circulatory and lymphatic systems in body maintenance. After you complete your study of this chapter, you will be able to:

- identify and locate the organs of the digestive system;
- describe the standard structure and functional aspects of the wall of the digestive tube;
- specify the contributions of lips, cheeks, teeth, and salivary glands to the digestive function;
- identify structures in the stomach, small intestine, and large intestine;
- explain the contributions of the liver and pancreas to the digestive process;
- indicate what materials are absorbed in the various organs of the digestive system.

The digestive system consists basically of a tube extending from the mouth to the anus. Along the tube are accessory glands. Passing through the system, usable portions of food are broken down, digested, and absorbed while rejected portions are expelled. In this chapter we shall examine the anatomy of the various portions of the digestive tube and see how they contribute to nourishing the human body.

ORGANS OF THE SYSTEM

1. The digestive system is made up of many organs, most of which make up portions of the digestive tube. Along this tube the food, called chyme, is processed mechanically and chemically. Then the usable portions are absorbed into the body fluids.

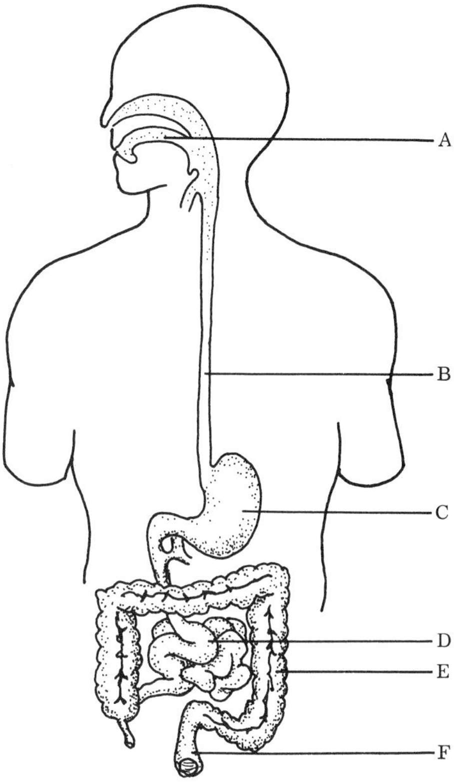

The first portion of the tube is the oral cavity (A in the diagram at the right), where food is introduced into the body. Food then passes into the esophagus (B), which transports it to the stomach (C). Most absorption of nutrients takes place in the small intestine (D), while additional water is absorbed in the large intestine (E). Finally, the unused remainder is passed into the rectum (F) for elimination.

(a) Which organ along the digestive tube is widest, or most dilated? ________________

(b) Which organ is longest?

____

(c) Arrange the organs listed below in the sequence food passes through them. The first in the sequence has been filled in.

________	large intestine	________	small intestine
________	stomach	________	rectum
1	oral cavity	________	esophagus

(d) Where is the pharynx located? ______________________________

- - - - - - - - - - - - - - - - - -

(a) stomach
(b) small intestine
(c) 5, 3, 1, 4, 6, 2
(d) between oral cavity and esophagus

2. In addition to the organs that form the digestive tube, the digestive system includes several major glandular organs. These are located outside the tube but empty their secretions into it. In the diagram on the following page, you can see the salivary glands, the liver, and the pancreas. The gall bladder is also outside the tube, and stores bile manufactured by the liver until it is needed by the digestive system.

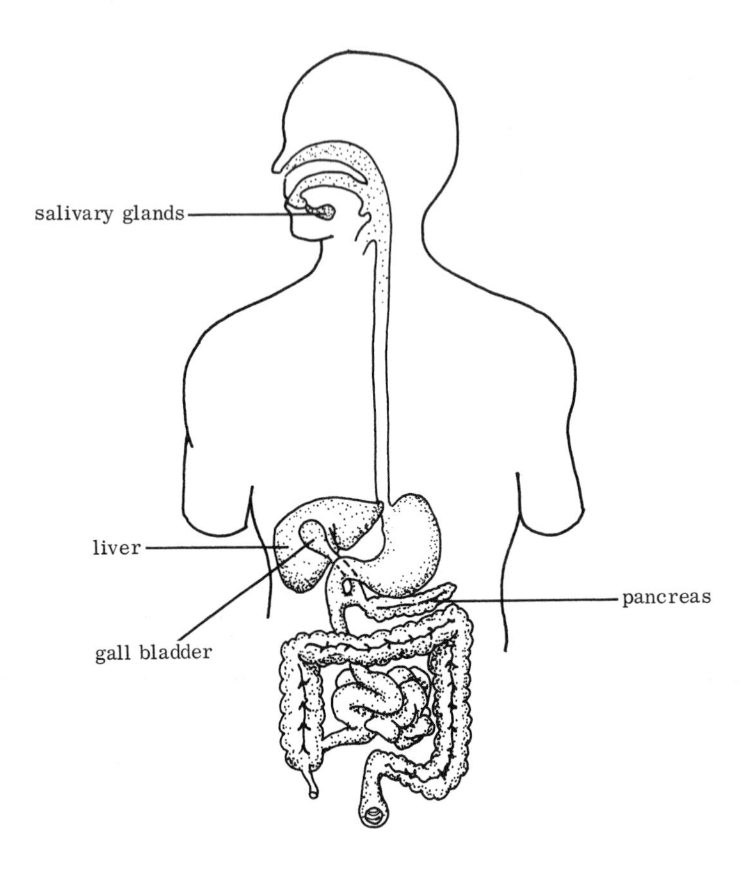

(a) Salivary glands empty into which region of the digestive system?

____

(b) Name the digestive gland located just inferior to the stomach.

____

(c) Anatomically, how would you describe the location of the liver?

____

(d) Which digestive organ is neither a portion of the digestive tube nor a gland? ____________________

(e) Into which organ of the digestive system do the liver and pancreas empty? ____________________

- - - - - - - - - - - - - - - - - - -

(a) oral cavity; (b) pancreas; (c) upper right abdomen; (d) gall bladder; (e) small intestine

The oral cavity

3. The mouth and oral cavity make up the first portion of the digestive tube. The lips and cheeks form the outer walls, while the palate forms the roof. The teeth break up ingested food, mixing it with saliva, while the tongue moves the food so that it is all at the mercy of the grinding teeth. The mechanical action of physically breaking food into smaller particles is the primary function of the mouth and oral cavity.

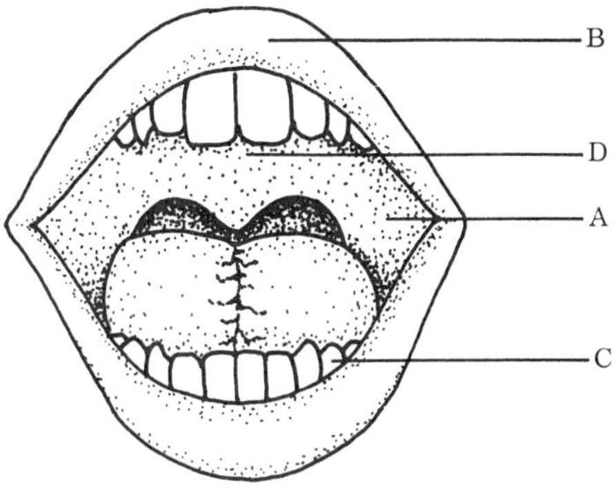

(a) Identify structure A in the diagram at the right.

____

How does it contribute to forming the mouth?

____

(b) Identify structure B.

____

Is its function more similar to that of A or C?

____

(c) What is a function of structure C? ______________________

(d) Identify structure D. ______________________

B
D
A
C

- - - - - - - - - - - - - - - - - -

(a) cheek; wall of oral cavity
(b) upper lip; more similar to cheek (A) than teeth
(c) grinding, breaking up food
(d) palate

4. The teeth are the most functional parts of the oral cavity. The 20 primary teeth of childhood and the 32 permanent teeth are all different. Both the upper jaw (maxillae) and the lower jaw (mandible) contain four basic types of teeth. The incisors in the very front of the mouth are used for biting or cutting food. The cuspid or canine teeth tear food. The bicuspids or premolars tear and crush food while the molars grind it.

Most adults have in each jaw four incisors, two cuspids, four premolars, and six molars. The last molar, on the left and right in each jaw, are the third molars or wisdom teeth; many people today never develop these molars.

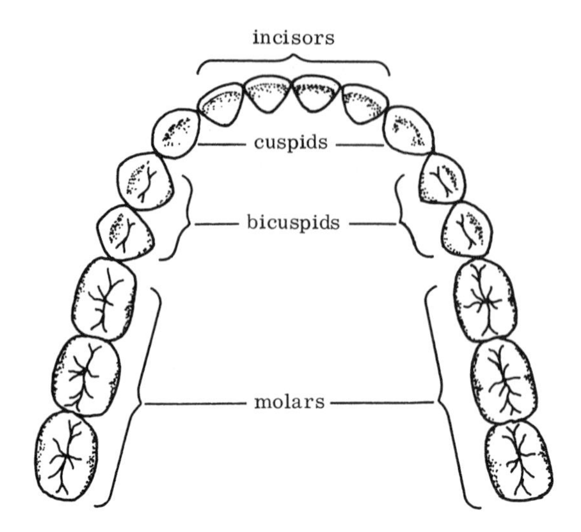

(a) Which teeth function in taking a bite from an apple? ____________

(b) Which teeth function in chewing a piece of steak? ______________

(c) How many mandibular teeth would a person have if only one of his lower third molars were present? ___________________

- - - - - - - - - - - - - - - - - -

(a) incisors; (b) molars; (c) 15

5. In the oral cavity both mechanical and chemical breakdown of food begins. The teeth contribute to the mechanical aspect, aided by muscles of the tongue, lips, cheeks, and jaws. The chemical contribution comes from the salivary glands. The process of chewing, or mastication, mechanically combines ingested food with the chemical secretions of salivary glands.

Three major pairs of salivary glands provide most of the saliva, which contains mucous, water, a protein enzyme, and small amounts of other substances. The enzyme, amylase, begins the chemical breakdown of starches, while the water and mucous secretions help to bring the food to a swallowable consistency.

The locations of the major salivary glands were shown in the diagram in frame 2. The major glands are located in pairs on opposite sides of the oral cavity region. The parotid gland is anterior to the lower border of the ear. The submandibular gland is near the angle of the mandible, and the sublingual salivary gland is beneath the anterior part of the tongue. Other minor salivary glands are scattered throughout the oral cavity and the mucous membrane lining it.

(a) What enzyme is found in saliva? ____________________ On what food group does this enzyme act? ____________________

(b) Name three constituents of saliva. ____________________

____

(c) Name the salivary glands indicated in the diagram.

A ____________________

B ____________________

C ____________________

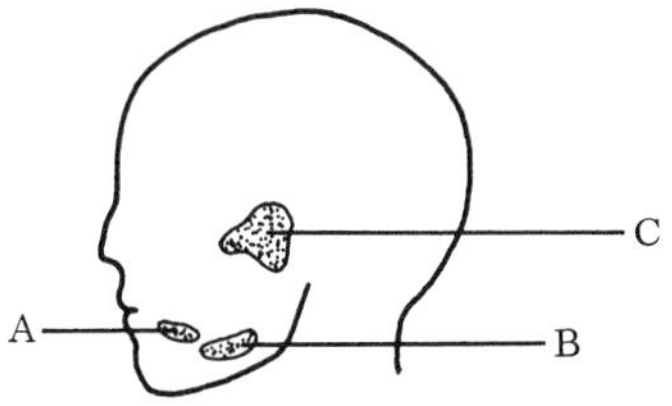

- - - - - - - - - - - - - - - - - -

(a) amylase; starches
(b) mucous, water, and protein or amylase
(c) A—sublingual; B—submandibular; C—parotid

6. After food is thoroughly masticated, it is swallowed. The technical term for swallowing is deglutition. The process of deglutition propels the food from the oral cavity into the oral pharynx. The action causes one of the laryngeal cartilages, the epiglottis, to seal off the entrance of the air passage to the lungs. This ensures that nutritive material remains within the digestive system and out of the respiratory system.

(a) Recall from the previous chapter the structure of the pharynx. What are its three parts? ____________________

____

(b) Write the anatomical words for the following processes.

chewing ____________________

swallowing ____________________

(c) Which laryngeal cartilage functions to close off the entrance to the larynx during swallowing? ____________________

- - - - - - - - - - - - - - - - - -

(a) nasal pharynx, oral pharynx, and laryngeal pharynx
(b) mastication; deglutition
(c) epiglottis

The upper digestive system

7. The oral cavity and pharynx lead the way into the digestive tube proper. The pharynx is also a portion of the respiratory system, and the oral cavity too can function in both systems. The wall of the remainder of the digestive tube has common elements, so this is a reasonable point to examine the basic features of the walls of the digestive system organs.

   The wall of a digestive organ has four layers. The inner layer is the mucous membrane, or mucosa; it includes the epithelium and a layer of connective tissue called the lamina propria. Two thin sheets of muscle tissue called the muscularis mucosa make up the outer edge of the mucosa. The submucosa layer is made of loose connective tissue with somewhat coarser fibers than the lamina propria. In the submucosa is a network of nerve fibers called Meissner's plexus, in addition to many blood vessels. The submucosa connects the muscularis mucosa to the external muscular layer. This third layer contains two sheets of smooth muscle, the inner one circularly oriented and the outer one longitudinally oriented. Between the two sheets of muscle is found another network of nerves, called the myenteric plexus (plexus means network). The outer layer or covering of the digestive tube is called the adventitia or serous layer (serosa) depending on its composition and location. While the exact structure of the four layers may vary in different organs, they are present in general form throughout the digestive system.

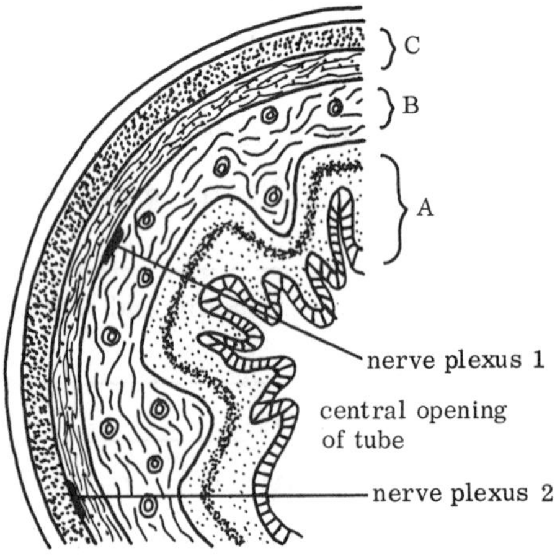

   (a) Name layer A in the diagram at the right. ______________
   From the inside, name its three components.

   ____

   ____

   ____

   (b) Name layer B ______________
   Identify the structure labeled nerve plexus 1. ______________

   ____

   (c) Give the orientation of each muscle sheet in layer C.

   inner ______________________ outer ______________________

   Identify nerve plexus 2. ______________________

   (d) Give two alternative names for the outer layer of the digestive tube wall. ______________________________

- - - - - - - - - - - - - - - - - - -

(a) mucosa; epithelium, lamina propria, and muscle
(b) submucosa; Meissner's plexus
(c) inner circular and outer longitudinal; myenteric plexus
(d) adventitia; serosa

8. Food passes from the pharynx into the esophagus, a straight tube about 25 cm long. The esophagus passes through the mediastinum (which separates the right and left thoracic cavities), just posterior to the trachea, and extends through an opening in the diaphragm.

The esophagus has the four layers in its wall. Its mucosa has stratified squamous epithelium, and a few glands are found in the lamina propria. The muscularis mucosa is quite well developed. The submucosa also contains a few esophageal glands. The external muscular layer shows a change from the pharynx to the stomach. About the first third of the esophagus has striated muscle, the middle third is a mixture of striated and smooth muscle, and the lower third is completely smooth muscle. The adventitia of the esophagus is made of fibrous connective tissue. Just after it passes through the diaphragm, the esophagus opens into the stomach, where thorough digestion begins.

(a) In the mucosa of the esophagus, the epithelial type is ______________ ______________. Are there glands in the lamina propria? ______________. Is the muscle layer well developed? ______________

(b) What kind of glands are found in the submucosa? ______________

(c) How could you distinguish microscopically between the upper, middle, and lower thirds of the esophagus? ______________ ______________

(d) What is the fibrous outer layer of the esophagus called? ______________ ______________

- - - - - - - - - - - - - - - - - - -

(a) stratified squamous; yes; yes
(b) esophageal
(c) the muscular layer differs—upper third is all striated, middle third is mixed, lower third is all smooth
(d) adventitia

9. The stomach is the first digestive organ that is completely within the abdominal cavity. It is connected to the esophagus in the midline, but extends to the left. The major gross anatomic features are shown in

the diagram below. The food, now called chyme, enters the stomach through the cardiac sphincter, so called because of its proximity to the heart. The chyme is accumulated in the fundus and body of the stomach, as its walls expand to accommodate the swallowed material. The folds of the inner walls of the stomach are called rugae (singular, ruga). The chyme is mixed with digestive substances secreted by the stomach and various gastric glands as it is funneled into the pyloric portion of the stomach. It is then passed through the pyloric sphincter into the small intestine.

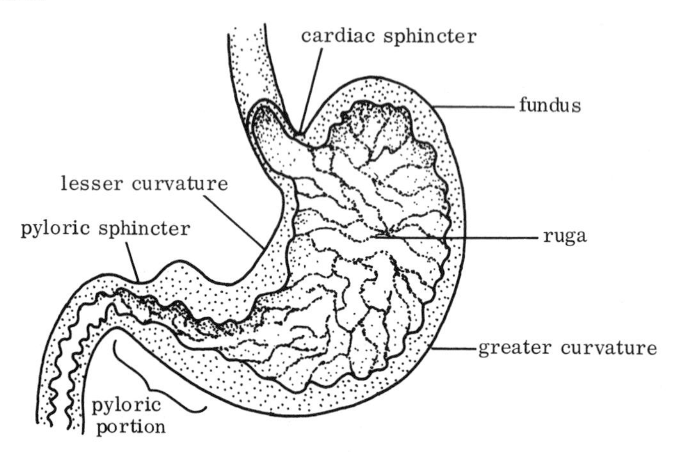

(a) Name the sphincters at either end of the stomach. ________________

____

(b) Which curvature of the stomach is more lateral than medial in the body? ________________________

(c) In which portion of the stomach does more mechanical action take place? ________________________

(d) What are the ridges inside the stomach called? ____________________

- - - - - - - - - - - - - - - - - - -

(a) cardiac and pyloric; (b) greater curvature; (c) pyloric portion; (d) rugae

10. The stomach contains the basic four-layered digestive wall, with several variations related to its function. The epithelium of the mucosa is simple columnar in type. Millions of simple tubular gastric glands are found in the mucosal layer, extending through the lamina propria. Several types of cells usually are found in each gland, but basically they secrete much mucous, hydrochloric acid, pepsin (which begins the

breakdown of proteins), and lipase (which works on fats). The submucosa also contains mucous glands. The muscular layer in the stomach contains a third sheet of smooth muscle on the inside; this one is obliquely oriented. The circular layer is thus the middle, while the outer longitudinal sheet remains as usual. The serosa of the stomach is part of the visceral peritoneum, which will be discussed later.

(a) Compare the epithelial lining of the stomach to that of the esophagus.

____

(b) Name three secretions of gastric glands. ____________________

____

(c) How is the external muscular layer of the stomach different from the lower third of the esophagus? ____________________

____

(d) What secretion of gastric glands begins the digestion of proteins?

____________________ What secretion causes "excess acid"?

____

- - - - - - - - - - - - - - - - - -

(a) stomach has simple columnar, while esophagus stratified squamous
(b) mucous, hydrochloric acid, pepsin, lipase (any three)
(c) stomach has an extra inner oblique layer of smooth muscle
(d) pepsin; hydrochloric acid

11. Absorption of nutrients begins in the stomach, but is not extensive. Small molecules such as alcohol or water may be absorbed directly from the stomach, as may some salts. Most ingested material, however, is stored in the fundus or body of the stomach, eventually moved to the pyloric portion for mixing with gastric secretions, then propelled into the small intestine.

(a) Which of the ingested materials below might be absorbed directly from the stomach?

__________ milk

__________ water

__________ scotch-on-the-rocks

__________ protein supplements

__________ chocolate bars

(b) Name at least three functions that take place in the stomach.______

____

- - - - - - - - - - - - - - - - - -

(a) water and scotch-on-the-rocks; (b) storage, mixing, secretion, and absorption

The small intestine

12. After it passes through the pyloric sphincter, chyme enters the small intestine, which is about 2.5 cm in diameter and 3 to 6 meters in length. The small intestine is divided into three sections. The duodenum forms a C-shaped portion about 25 cm long. The liver and pancreas have ducts which empty into the duodenum. The jejunum is the next segment of small intestine, while the ileum is the last, joining the small intestine to the large intestine. These three portions cannot be distinguished except microscopically. The small intestine is arranged in coils, filling most of the abdominal cavity.

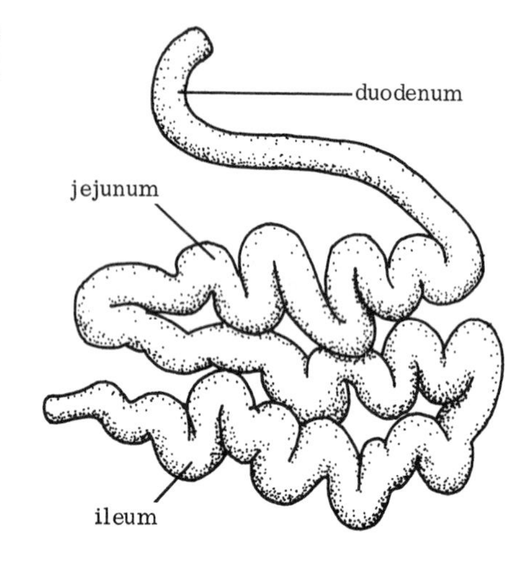

(a) Into which segment of the small intestine does the pyloric sphincter open? ____________________

(b) Which segment of the small intestine joins it to the large intestine? ____________________

(c) Into which segment of the small intestine do the liver and pancreas empty? ____________________

(d) What small intestine segment leads into the jejunum? ____________ What segment leads out of the jejunum? ____________________

- - - - - - - - - - - - - - - - - -

(a) duodenum
(b) ileum
(c) duodenum
(d) duodenum; ileum

13. The wall of the small intestine has the four usual layers, with certain specializations for its primary purpose—absorption. One specialization, plicae circulares (circular folds), involves both the mucosa and the submucosa. These plicae circulares are circular ridges in the intestinal

lining, and are most prominent in the second half of the duodenum and the first half of the jejunum; they greatly increase the surface area for absorption. The intestinal villi, another specialization, are outgrowths of the mucosa. These fingerlike, epithelial-covered structures contain lacteals (lymph capillaries) and blood capillaries that absorb nutrients. The absorptive columnar epithelial cells of the small intestine contain microvilli on their free ends, which also vastly increase the surface area. Goblet cells are common, secreting mucous to lubricate the passage of chyme. Other glands in the mucous layer of the small intestine contribute digestive enzymes that supplement the contributions of the liver and pancreas.

(a) What specialization for absorption involves both the mucosal and submucosal layers of the small intestine? ______________

(b) Which specialization for absorption involves only epithelial cells?

____

(c) Which specialization for absorption houses lacteals and blood capillaries? ______________

- - - - - - - - - - - - - - - - - -

(a) plicae circulares (circular folds); (b) microvilli; (c) intestinal villi

14. As mentioned in frame 13, the mucosa and submucosa of the small intestine are specialized for absorption, in addition to containing many glands. The muscular layer contains the typical two-layer structure, with a nerve plexus between. The serosa, as in the stomach, is composed of visceral peritoneum. Mesenteric arteries and veins supply the small intestine with most of the nutrient-laden blood being diverted to the liver via the portal vein.

(a) Is the muscular layer of the small intestine more similar to that of the stomach or that of the esophagus? ______________

(b) Is the outer layer of the small intestinal wall more similar to that of the stomach or esophagus? ______________

(c) Which vein carries nutrient-laden blood from the small intestine to the liver? ______________

- - - - - - - - - - - - - - - - - -

(a) esophagus (2 layers); (b) stomach (serous, visceral peritoneum); (c) portal vein

15. Most of digestion and absorption takes place in the small intestine, after the addition of liver and pancreatic secretions. Water is absorbed at all levels of the small intestine. Carbohydrates (monosaccharides) are absorbed into the portal blood system and taken directly to the liver. Fats and proteins are absorbed into lacteals and carried to the thoracic duct, where they join the circulatory system. We will cover the specific contributions of liver and pancreatic enzymes a bit later in this chapter.

    (a) In which segments of the small intestine is water absorbed? ______

    ____

    (b) What components of food are absorbed into lacteals? ____________

    ____

    (c) What component of food is absorbed directly into the portal system?

    ____

- - - - - - - - - - - - - - - - - - - -

(a) duodenum, jejunum, and ileum (that is, in all segments); (b) proteins and fats; (c) carbohydrates

The lower digestive system

16. The large intestine is about 6 cm in diameter and about 1.5 meters long. It is divided into the cecum, the colon, the rectum, and the anal canal.

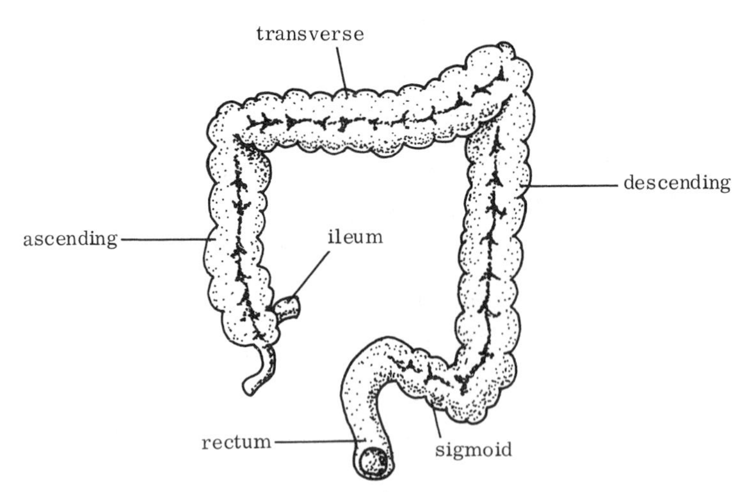

The ileum of the small intestine is joined to the large intestine at the ileo-cecal valve. The cecum is a blind pouch below the valve, with another blind tube, the vermiform appendix, below it. The colon has four portions: ascending, transverse, descending, and sigmoid (S-shaped). The rectum extends from the sigmoid colon along the sacrum and coccyx for about 15 cm to the pelvic floor. At this point, the anal canal passes between the levator ani muscles to the anus.

(a) The junction of which two segments of the colon is nearest the liver?

____

(b) The junction of which two segments of the colon is nearest the stomach? ______________________

(c) Is the appendix part of the small or large intestine? ____________

(d) Name the valve that connects the small and large intestines. ______

____

(e) What three segments connect the descending colon to the anus?

____

- - - - - - - - - - - - - - - - - -

(a) ascending and transverse; (b) transverse and descending; (c) large; (d) ileo-cecal valve; (e) sigmoid colon, rectum, and anal canal

17. The epithelium of the large intestine is simple columnar with no microvilli. Goblet cells and mucosal glands secrete mucous. The submucosa is similar to that in the small intestine, although there are no plicae circulares in the large intestine. The muscular layer differs in its longitudinal sheet in the large intestine. The longitudinal sheet is arranged in three strips called taeniae coli. These are shorter than the intestine, and cause it to be gathered up into sacs called haustra. The serosa, as in most of the abdominal organs, consists of visceral peritoneum. In the anal canal, the epithelium changes to stratified squamous. No haustra are present in the rectum or anal canal, where rectal veins are close to the surface. Bacteria are numerous in the large intestine, and much water is absorbed from the contents.

(a) In what other organ of the digestive system is the epithelial lining similar to that in the anal canal? ______________________________

(b) What layer creates the three taeniae coli of the colon? __________

____

(c) What are the sacs created by taeniae coli called? ______________

(d) What is absorbed from colonic contents? ______________________

- - - - - - - - - - - - - - - - - -

(a) esophagus; (b) outer longitudinal sheet of muscular layer; (c) haustra; (d) water

18. The abdominal cavity is lined by a large serous membrane made up mesothelium (a form of simple squamous epithelium) and connective tissue. This membrane is called the peritoneum, and is double-layered, having a parietal and visceral layer, as in the pleura of the thoracic cavity. The parietal peritoneum is closely attached to the abdominal wall. The visceral peritoneum covers most of the abdominal organs. Extensions of the visceral peritoneum make up the omentum, which covers the stomach, and the mesentery, which supports and surrounds the small intestine.

    (a) Name the serous lining of the abdominal cavity. ________________

    (b) What makes up the serosa of the large intestine? ________________

    (c) What extension of the peritoneum covers the stomach? ____________ ____________________

    (d) What supports the small intestine in coils in the abdominal cavity? ____________________

    (e) Would you expect the anal canal to be covered with visceral peritoneum? ______________________ Why? ______________________

- - - - - - - - - - - - - - - - - -

(a) parietal peritoneum; (b) visceral peritoneum; (c) omentum; (d) mesentery; (e) no, because it's in the pelvic floor, not the abdominal cavity

Accessory digestive glands

19. The liver is located outside the digestive tube, but connects with it by a duct. It is located in the upper right quadrant of the abdomen, with its upper surface molded by the diaphragm. The liver has two main lobes, the right and left, with two smaller lobes, the caudate and quadate, attached to the larger right lobe of the liver. Most of the surface of the liver is covered with peritoneum overlying a fibrous capsule on the liver itself.

    (a) Where is the liver located? ______________________________

    (b) Name the four lobes of the liver. ____________________________ ______________________

(c) What muscle determines the shape of the superior surface of the liver? ____________________

- - - - - - - - - - - - - - - - - -

(a) upper right abdomen; (b) right, left, caudate, and quadrate;
(c) diaphragm

20. At one point on the capsule of the liver (hepatic capsule), the hepatic artery and portal vein enter a connective tissue trunk that then branches out to supply all parts of the liver. Lymphatics and bile ducts also follow the connective tissue tree. The blood vessels subdivide until they produce capillaries around all the liver cells. Sinusoids, or widened blood spaces, are present between hepatic lobules, where worn out blood cells may be destroyed or toxic substances removed from the blood.

The hepatic lobules are the functional units of the liver. Here bile is produced and blood detoxified. The hepatic vein drains the liver into the vena cava. The primary digestive function of the liver is the production of bile. This substance, vital to the digestion of fats, is manufactured in liver cells, then secreted into bile capillaries that are adjacent to each cell. The bile capillaries unite and merge, following the pathway of the blood vessels. Eventually two hepatic ducts leave the liver, then merge into the common bile duct, which empties into the duodenum. A branch from the common bile duct, the cystic duct, leads to the gall bladder where bile may be stored until needed in the small intestine.

(a) Name the duct branches indicated in the diagram at the right.

A ____________________

B ____________________

C ____________________

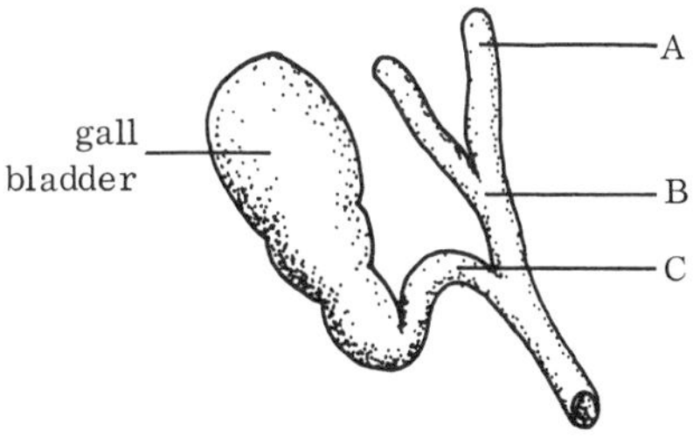

gall bladder

A

B

C

(b) What is the primary digestive function of the liver?

____

(c) What happens in the sinusoids of the liver? ____________________

____

- - - - - - - - - - - - - - - - - -

(a) A—hepatic duct; B—common bile duct; C—cystic duct; (b) produce bile to digest fats; (c) blood cells destroyed, materials detoxified

21. The pancreas lies with its tail and body behind the stomach, while its head is close to the curve of the duodenum. The endocrine portion of the pancreas secretes insulin and glucagon and functions in the utilization of glucose by the body as was discussed in Chapter 7. The exocrine portion of the pancreas secretes pancreatic juice, which includes several different enzymes. The various enzymes of the pancreas act on all the food types—fats, carbohydrates, and proteins. The pancreatic juice empties into the pancreatic duct, which enters the duodenum with the common bile duct.

   (a) What is secreted by the endocrine pancreas. ______________________

   (b) On what food groups do secretions of the exocrine pancreas act?

   ____

   (c) Into what organ of the digestive system does pancreatic juice empty?

   ____

- - - - - - - - - - - - - - - - - -

(a) insulin and glucagon; (b) all (fats, carbohydrates, proteins);
(c) small intestine (duodenum)

COMMON DISORDERS OF THE DIGESTIVE SYSTEM

Dental caries, or cavities

Cavities in the teeth seem to result in part from overeating sweet foods. The acids produced when sweets are broken down in the mouth tend to weaken the enamel and its resistance to bacteria. The condition can cause pain and interfere with proper chewing of food.

Ulcers

Ulcers in the stomach and duodenum seem to result from some interaction of a stressful situation and overproduction of acid in the stomach. The result can be painful and result in bleeding if the ulcer penetrates blood vessels. Removal of portions of the stomach and duodenum in ulcer surgery, can, of course, interfere with the digestive process.

Appendicitis

Since the appendix is a blind tube, it can easily accumulate bacteria and food debris. If its open end should become blocked, inflammation will result, causing nausea, vomiting, pain, and possibly eventual death.

Removal of the appendix to prevent rupture and general peritonitis is the procedure of choice in appendicitis. The surgical procedure has only a transient effect on the digestive tube.

Cirrhosis of the liver

This condition, usually precipitated by malnutrition and excessive alcohol intake, shows a decreased number of functioning liver cells. Since the liver functions in a wider variety of activities than any other organ, cirrhosis can be extremely debilitating.

SELF-TEST

This Self-Test is designed to show how well you have mastered this chapter's objectives. Answer each question to the best of your ability. Correct answers and review instructions are given at the end of the test.

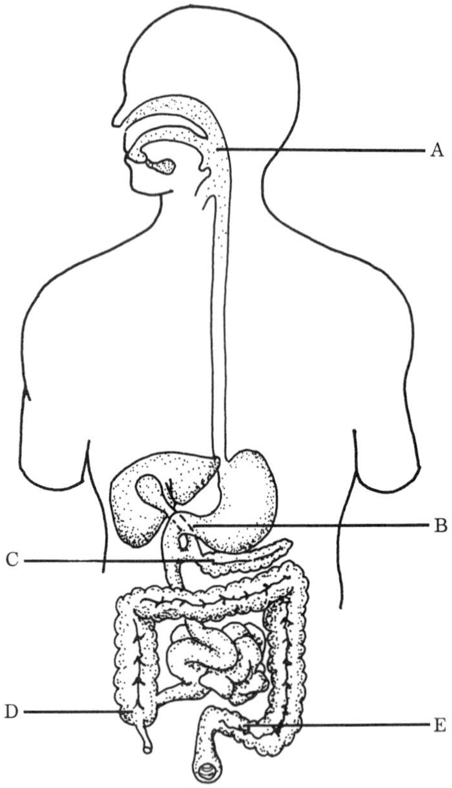

1. Identify the organs and parts of the digestive tube indicated on the right.

   A ____________________

   B ____________________

   C ____________________

   D ____________________

   E ____________________

2. Describe the muscular layer of the wall of the large intestine.

   ____

   ____

   ____

3. What type of epithelium is found in the esophagus?

   ____

4. Name two nerve plexuses that might be found in the wall of the small intestine, and name the layer in which each is found.

   ____

   ____

5. How does the muscular layer of the stomach wall differ from that of the esophagus? ____________________

6. In which digestive organ does the major part of digestion and absorption take place? ____________________

7. What structure forms the serosa of abdominal digestive organs? ____________________

   ____

8. What is the function of the digestive contribution of the liver? ________

____

9. Into what part of the digestive tube does the pancreas release its exocrine secretion? ____

10. Name three specializations in the small intestine that increase the surface area for absorption. ______________________________________

____

Answers

Compare your answers to the Self-Test questions with those answers given below. If all of your answers are correct, you are ready to go on to the next chapter. If you missed any, review the frames indicated in parentheses following the answers. If you missed several questions, you should probably reread the entire chapter carefully.

1. A—oral pharynx; B—pyloric portion (stomach); C—head (pancreas); D—cecum (large intestine); E—sigmoid colon (large intestine) (frames 1, 9, and 16)
2. inner circular layer, outer longitudinal layer is three flat bands called taeniae coli (frame 17)
3. stratified squamous (frame 8)
4. Meissner's in the submucosa; myenteric in the muscular layer (frame 7)
5. stomach has a third layer, oblique, on the inside (frames 8 and 10)
6. small intestine (frame 13)
7. visceral peritoneum (frame 18)
8. digest fats (frame 20)
9. duodenum (frame 21)
10. plicae circulares, villi, and microvilli (frame 13)

CHAPTER TWELVE
Skin and Associated Structures

The skin and its associated organs play an important role in the maintenance of the body. The most obvious function is that of protection, but prevention of fluid loss and temperature regulation are also provided by the integumentary system. After you complete your study of this chapter, you will be able to:

- identify the three parts of skin as epidermis, dermis, and hypodermis;
- differentiate among the five layers of epidermis, stating distinguishing features of each layer;
- describe the epithelium/connective tissue interface in the skin;
- specify the components of the two layers of the dermis;
- identify the components of the hypodermis;
- explain how the skin functions;
- identify the various components of a hair and its follicle;
- contrast the location and secretions of sebaceous and sweat glands;
- describe the development of a fingernail or toenail.

Skin performs its functions by virtue of its structure; its cells are so closely attached that it is virtually waterproof. Even bacteria cannot pass through unbroken skin.

THE EPIDERMIS

1. Skin is composed of three major layers: the epidermis, the dermis, and the hypodermis. The epidermis is located on the free surface of the skin. It is composed of keratinized stratified squamous epithelium. The five layers of epidermis are shown in the diagram on the following page.

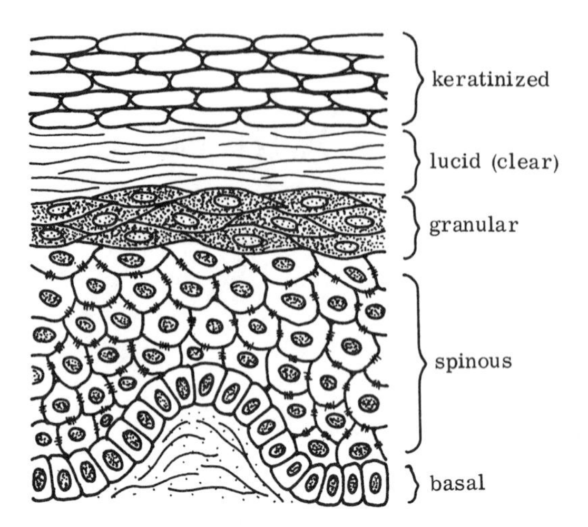

At the base of the epithelium the layer of cuboidal cells is called the basal layer. The cells in the basal layer and the adjacent spinous layer are often called the germinating layer, because cell division or mitosis occurs here.

The cells in the spinous layer show what look like spines or prickles. These are actually fibers from cell connections called desmosomes. The desmosomes tightly attach the cells of the spinous layer together, as well as to the basal and granular layers. The desmosomes are not "intercellular bridges," as they once were called, because no components pass from one cell into an adjacent cell. A diagram of a desmosome shows that the cells maintain separate identities. The cells in the granular layer have fewer desmosomes, but do contain many granules. The granules increase in number as other cellular components decrease. The cells in the lucid layer are clear and rather structureless, while the keratinized cells on the surface are scaly and can be lost with abrasion.

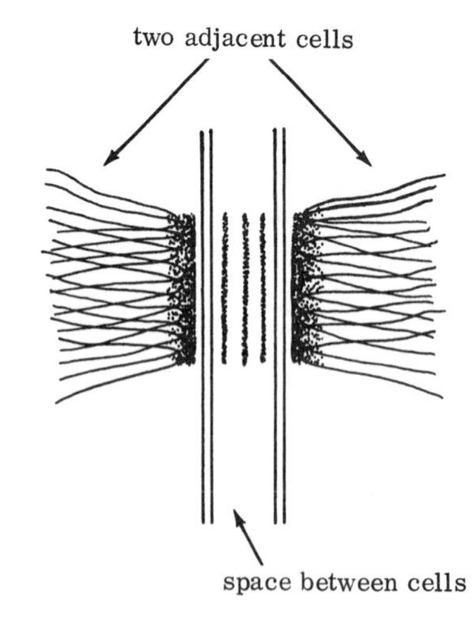

(a) Which layer of the epidermis contains more desmosomes?

____

(b) Which layer of the epidermis is clear? ____________________

(c) Name the innermost two layers of the epidermis. ____________________

(d) Name the layer of the epidermis found between the clear layer and the layer that has the most desmosomes. ______________

(e) Cells of which layer rub off when the skin is abraded? ______________

- - - - - - - - - - - - - - - - - -

(a) spinous; (b) lucid; (c) basal and spinous; (d) granular; (d) keratinized

2. The outer two layers of the keratinized epithelium of the skin are fairly similar, except that the lucid layer contains a substance called eleidin, which gives it a clear appearance. As the eleidin is converted into keratin, the cells attain the scaly quality of the keratinized layer. The keratinization process is already under way in the middle layer, for the granules contain a protein called keratohyalin, which forms eleidin.

(a) Name the layers of the epidermis indicated in the figure at the right.

A ______________

B ______________

C ______________

D ______________

E ______________

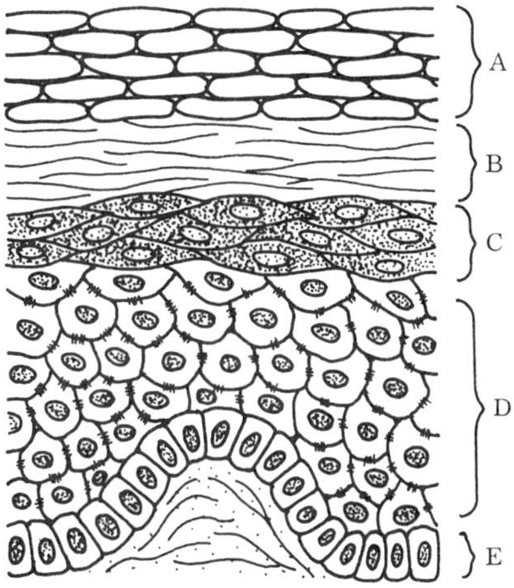

(b) List one distinguishing characteristic of each layer.

A ______________

B ______________

C ______________

D ______________

E ______________

(c) In which layer(s) of epidermis does cell division take place? ______________

- - - - - - - - - - - - - - - - - -

(a) A—keratinized; B—lucid; C—granular; D—spinous; E—basal
(b) A—scaly; B—clear; C—granules; D—desmosomes; E—columnar

(or other characteristics from frame 1 or 2)
(c) basal and spinous (germinating)

The dermis

3. The second layer of skin is the dermis, which underlies the epidermis. The basal layer of epidermis is arranged in folds; and the outer layer of the dermis has pegs or papillae which interdigitate with these folds. For this reason the more superficial layer of dermis is called the papillary layer. It contains a thin, fibroelastic connective tissue.

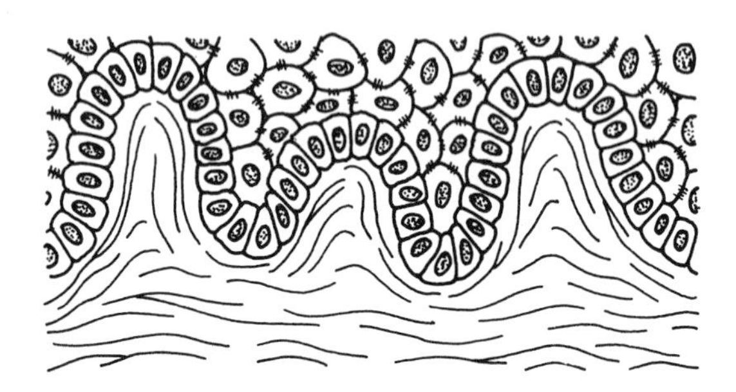

The remainder of the dermis forms the reticular layer, about five times as thick as the papillary layer. The reticular layer is composed of a coarser and denser fibroelastic connective tissue. Derivatives of the skin, such as hair and sweat glands, are found in the dermis, as are blood vessels and nerves at the interface of the epidermis and dermis. Capillaries extend into the connective tissue papillae capillaries but do not penetrate the basal layer of the epidermis. Nerve endings, however, often penetrate into the epidermis.

(a) Name the two layers of the dermis. ______________________________

(b) No blood vessels are present in epithelium. Where do basal cells get their nourishment? ______________________

- - - - - - - - - - - - - - - - - -

(a) papillary and reticular; (b) papillary capillaries

4. The four main functions of the skin are protection, thermoregulation, secretion, and sensation. The closely adherent cells and scaly outer layer of skin provide for the protective function. Thermoregulation is provided to some extent by the papillary capillaries that dilate or constrict depending upon temperature needs. Sweat glands of the skin also aid in cooling the body, for evaporation of the watery sweat removes body heat. Thus the secretory and thermoregulative functions are both met by sweat glands. Sebaceous glands, one of which is associated

with each hair follicle, secrete an oily substance that lubricates the hair and surrounding skin, thus also contributing to secretory functions. The nerve endings in the epidermis provide for sensation. These receptors are specific for pain, touch, pressure, and temperature, as we discussed in Chapter 6.

(a) What feature of the skin contributes most to its protective function?

____

(b) Name components that contribute to the thermoregulative function of the skin. ______________________________

(c) Name two glands in the skin. What type of secretion does each have?

____

(d) Name three of the four sensations perceived through nerve endings in the skin. ______________________________

- - - - - - - - - - - - - - - - - -

(a) tightly connected cells
(b) papillary capillaries and sweat glands
(c) sweat gland—watery secretion; sebaceous gland—oily secretion
(d) pain, touch, pressure, and temperature (any three)

The hypodermis

5. The skin is connected to underlying tissues, such as muscles, by a layer of fatty connective tissue called the hypodermis. Injections are often given subcutaneously (under the skin) into the hypodermis, since this layer can hold much fluid. The base of sweat glands are located in the hypodermis.

Each sweat gland is a tubular structure with the secretory portion coiled beneath the dermis. The excretory duct extends to the free surface of the epidermis, and is lined with epithelial cells. Sweat contains water and many of the organic constituents of blood and tissue fluid. It is manufactured in cells, and released into ducts. Its chief salt is sodium chloride (NaCl). The sweaty secretion is generally stimulated by action of increased blood temperature on centers in the brain, and it results in cooling by evaporation. Sweating also contributes to the excretory function of the body in that such elements as ammonia and uric acid may be removed from the body in this way.

(a) What portion of a sweat gland is found in the dermis? ____________

____

(b) Where, anatomically, are the secretory portions of the sweat glands found? ____________

(c) What process enables secretions of sweat glands to function in thermoregulation? ____________

- - - - - - - - - - - - - - - - - - -

(a) part of the excretory duct; (b) in the hypodermis; (c) evaporation

ASSOCIATED STRUCTURES

6. The sebaceous gland of the skin is always associated with a hair follicle. In fact, it develops from the follicle. The secretion of a sebaceous gland is actually whole cells containing an oily material called sebum. The sebaceous glands are located within the reticular layer of the dermis, and empty into the hair follicle.

   (a) How does the secretion of a sebaceous gland differ from that of a sweat gland in consistency? ____________

   (b) In type of secretion? ____________

   (c) In location of duct opening? ____________

   (d) In which layer of the skin are sebaceous glands found? ____________

- - - - - - - - - - - - - - - - - - -

(a) sebaceous—oily; sweat—watery
(b) sebaceous—whole cells; sweat—produced by cells
(c) sebaceous—along hair shaft; sweat—on epidermis
(d) reticular layer of dermis

7. Hairs develop from the epidermis over most of the body surface. A deep epithelial and connective tissue sheath forms the hair follicle which grows down into the subcutaneous region. The root of a hair is at the base of the hair follicle. The epithelial cells at the base of the follicle are responsible for growth of hairs.

   The hair shaft has two parts in cross section. Cells of the central medullary core contain air spaces, while the cortex is keratinized. All of the hair shaft above the skin level is made up of dead, keratinized cells. A small arrector pili muscle is attached to each hair follicle; contraction of these muscles give rise to "goose bumps" on the skin.

A sebaceous gland is associated with each hair and contraction of the arrector pili muscle may also aid in expelling its secretions.

(a) Name the structures indicated at the right.

A ______________________

B ______________________

C ______________________

(b) Does a normal haircut affect the growth of hair? ______________

(c) Of what tissues does a hair follicle consist? ______________________

____

A

B

C

- - - - - - - - - - - - - - - - - - -

(a) A—sebaceous gland; B--arrector pili muscle; C—root of hair shaft or base of follicle
(b) no
(c) epithelial and connective tissue

8. Fingernails and toenails are modifications of the outer layer of the epidermis. They are made of hard keratin. At the root of each nail, air may be mixed with the keratin matrix, producing a white crescent called the lunula. The nail plate grows from this region and slides slowly over the nail bed, which corresponds to the deeper three layers of the epidermis. The pink color of nails results from underlying dermal capillaries.

(a) What produces the whitish crescent at each nail root? ____________

____

(b) What is the crescent called? ______________________

(c) How is the composition of nails similar to that of hair? ___________

____

(d) To what layers of epidermis does a nail bed correspond? __________

____

- - - - - - - - - - - - - - - - - - -

(a) air mixed with keratin; (b) lunula; (c) both are keratin; (d) basal, spinous, and granular

9. Suppose you fall on a board with a nail on it. We will examine (in slow motion) the path of the wound.

   (a) The nail passes through the epidermis and dermis. Name the seven layers in order. ______________________________

   ____

   (b) Would you be able to feel pain before the nail reached the dermis? ____________ Would it bleed before the nail reached the dermis? ____________

   (c) The nail cuts a muscle and a gland next to a hair follicle. Name the muscle and gland. ____________

   (d) The nail penetrates as far as the tissue directly beneath the dermis. What is it? ____________

- - - - - - - - - - - - - - - - - -

(a) keratinized, lucid, granular, spinous, basal, papillary, and reticular
(b) yes; no (nerve endings are in epidermis but not blood vessels)
(c) arrector pili muscle and sebaceous gland
(d) hypodermis

SELF-TEST

This Self-Test is designed to show how well you have mastered this chapter's objectives. Answer each question to the best of your ability. Correct answers and review instructions are given at the end of the test.

1. Which layer of the skin has the greatest number of cells held tightly together by desmosomes? ____________________
2. How do the deepest layer of epithelium and the most superficial layer of the connective tissue in the skin fit together? ____________________
3. In which layer of the skin are the most superficial capillaries found? ____________________
4. Where is the functional or secretory portion of a sweat gland located? ____________________
5. Which two layers of the epidermis are sometimes called the germinating layer? ____________________
6. The nail corresponds to which layers of the epidermis? ____________________
7. Which gland has entire cells as its secretion? ____________________
8. Which of the accessory structures discussed is found in the reticular layer of the dermis? ____________________
9. Give two aspects of the skin that contribute to thermoregulation. ____________________
10. Name three more functions of the skin. ____________________

Answers

Compare your answers to the Self-Test questions with those answers given below. If all of your answers are correct, you are ready to go on to the next chapter. If you missed any, review the frames indicated in parentheses following the answers. If you missed several questions, you should probably reread the entire chapter carefully.

1. The spinous layer has the most desmosomes (frame 1)
2. folds or projections (pegs or papillae) are interdigitated, holding tightly together (frame 3)
3. papillary (frame 3)
4. hypodermis (frame 5)
5. basal and spinous layers (frame 1)
6. lucid and keratinized (frame 8)
7. sebaceous gland (frame 6)
8. sebaceous gland, duct of sweat gland, hair follicle (any one) (frames 3 and 5)
9. sweat is evaporated; papillary capillaries constrict or dilate (frame 4)
10. protective, sensory, excretory, secretory (any three) (frame 4)

CHAPTER THIRTEEN

The Excretory System

The excretory system aids in maintaining the normal environment of the body by removing waste material. While several organs are involved in this process, this chapter will deal primarily with those in the urinary system. After you complete your study of this chapter, you will be able to:

- specify the materials excreted by the lungs, skin, and large intestine;
- identify and locate the gross structural organs of the urinary system;
- identify the following features of the kidney: cortex, medullary pyramid, papilla, calyces, hilus, and capsule;
- describe the structure and function of a nephron and its associated capillaries;
- explain the function of a glomerulus;
- list two steps in the formation of urine from blood;
- describe the epithelial lining of kidney tubules;
- describe the epithelial lining of the ureters, bladder, and urethra.

One of the critical functions of any organism is elimination of waste material. In the human organism, waste material includes such items as carbon dioxide, excess digestive juices, and undigested food. By-products are produced in many of the chemical reactions that take place within cells; these by-products must be removed in some way, as must over-supply of elements needed by the body. The urinary system, which produces urine from elements in the blood, performs most of the excretory functions, and is the primary topic of this chapter. But a brief review of the excretory function of the respiratory system, digestive system, and integumentary system is in order.

1. The respiratory system is responsible for the excretion of carbon dioxide wastes from the body. In the lungs, carbon dioxide released from the blood, along with water vapor and body heat, combines with

expired air. More water and heat are among the substances excreted through the skin, along with salts and other components of sweat. The large intestine passes along undigested food remains, excess digestive enzymes, and some more water.

(a) What is the primary substance disposed of through the skin? ______

____

(b) What is the excretory function of the lungs? ______________________

(c) How does the large intestine contribute to excretion? ____________

____

- - - - - - - - - - - - - - - - - - -

(a) water; (b) remove carbon dioxide from blood; (c) excretes left-overs after digestion

THE KIDNEYS

2. Since the blood is the principal means of transport of substances in the body, most of the waste material for excretion accumulates there. The urinary system has the responsibility for combining the waste material into urine and eliminating it.

The organs of the urinary system are shown at the right. Two bean-shaped kidneys are just above the waistline in the back of the abdominal cavity. In fact, they are retroperitoneal, behind the peritoneum. A ureter leads from each kidney to the urinary bladder, where urine is stored. The urethra, which differs between the sexes, conducts urine from the bladder to outside the body.

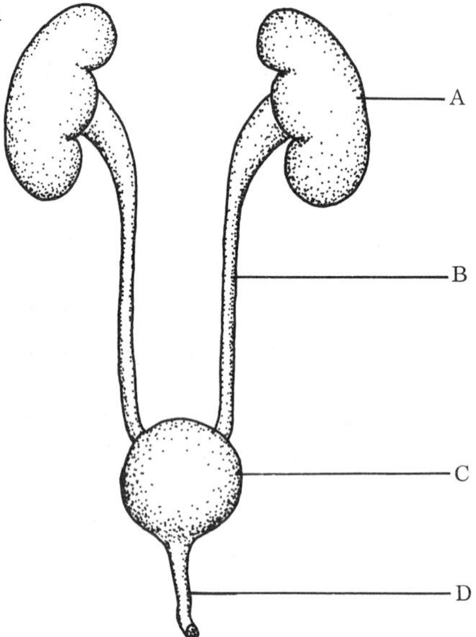

(a) How many organs make up the urinary system? ______

(b) What organs conduct urine from the kidneys to the bladder? ______________

(c) What organ of the urinary system communicates with the outside environment? ____________

(d) Label the structures indicated in the drawing.

A ____________________

B ____________________

C ____________________

D ____________________

- - - - - - - - - - - - - - - - - - -

(a) 6 (2 kidneys, 2 ureters, 1 bladder, 1 urethra)
(b) ureters
(c) urethra
(d) A—kidney; B—ureter; C—bladder; D—urethra

3. The kidneys are the functional organs of the urinary system, as the others are concerned with conducting or storing the prepared urine. Structurally, each kidney shows a medial concave surface and a lateral concave surface. The medial concave surface is known as the renal hilum. (The term renal refers to the kidney.) At the hilum, the renal artery and vein enter the kidney, along with the expanded upper end of the ureter, which is called the renal pelvis.

A thin fibrous capsule covers the kidney, and it is held in place next to the muscle of the posterior wall of the abdominal cavity by surrounding fatty tissue. The peritoneum is anterior to the kidneys. The adrenal glands (also called supra-renal glands) cap the kidneys, as you learned in Chapter 7. The kidneys and adrenal glands move slightly with respiration since they are directly beneath the diaphragm.

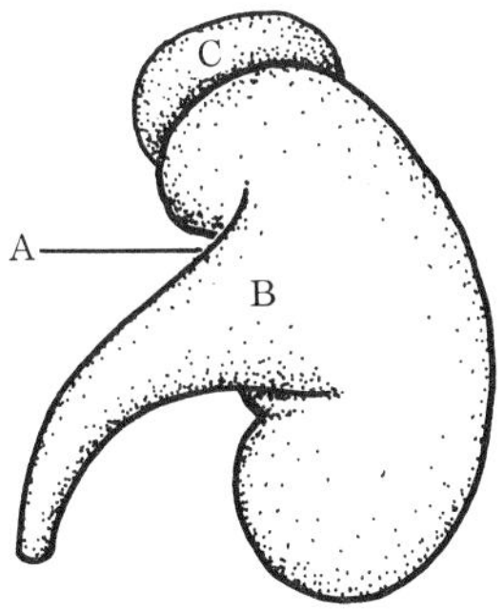

(a) In the diagram at the right, what is area A of the kidney called? ____________________

(b) Name structure B. ____________________

(c) Name structure C. ____________________

(d) Does the drawing represent a right or left kidney? ____________________

(e) Where is the kidney located with respect to:

the diaphragm ____________________

the peritoneum ____________________

the waistline ____________________

- - - - - - - - - - - - - - - - - - -

(a) hilum
(b) pelvis (expanded end of ureter)
(c) adrenal gland
(d) left
(e) below the diaphragm; behind the peritoneum; above the waistline

4. In cross section, functional features of the kidney can be seen with the naked eye. The cortex and medulla are shown in the following diagram. Medullary pyramids terminate in apices or papillae (singular, apex or papilla). These empty into funnel-like calyces (singular, calyx) of the pelvis.

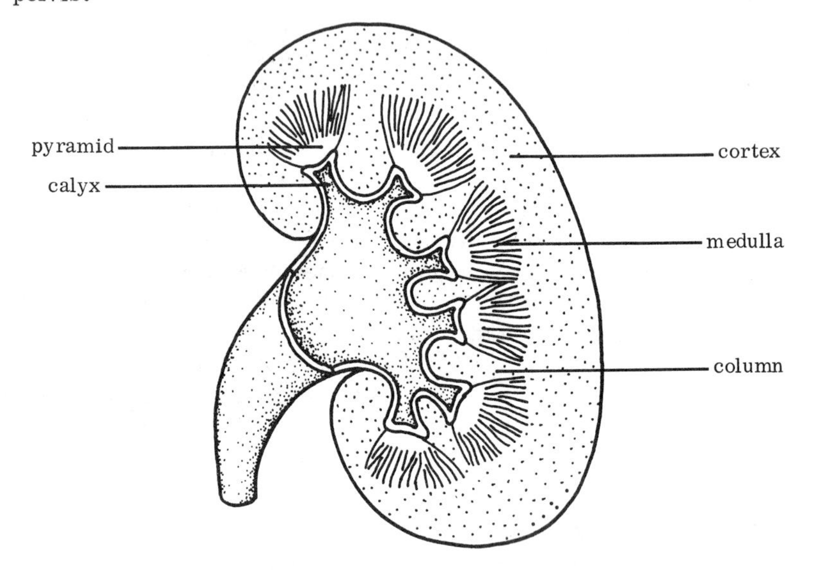

The outer portion of the kidney is the cortex, which extends in columns between the medullary pyramids.

These divisions reflect the microscopic structures located in various areas of the kidney. Each kidney contains ten or twelve pyramids; a corresponding number of calyces open into each pelvis, as each papilla empties into a different calyx.

(a) Indicate whether each structural feature listed below is a part of the renal cortex or medulla.

pyramid ______________________

column ______________________

outer portion of kidney ______________________

papilla ______________________

calyx ______________________

(b) Is the base of a medullary pyramid closer to the renal capsule or to the hilum? ______________________________

- - - - - - - - - - - - - - - - - - -

(a) pyramid—medulla; column—cortex; outer portion—cortex; papilla—medulla; calyx is not part of the kidney, but of the ureter
(b) capsule (apices of pyramids are nearer hilum)

5. On the diagram, label the following structures.

   A—hilum
   B—calyx
   C—pelvis
   D—medullary pyramid
   E—papilla
   F—renal column

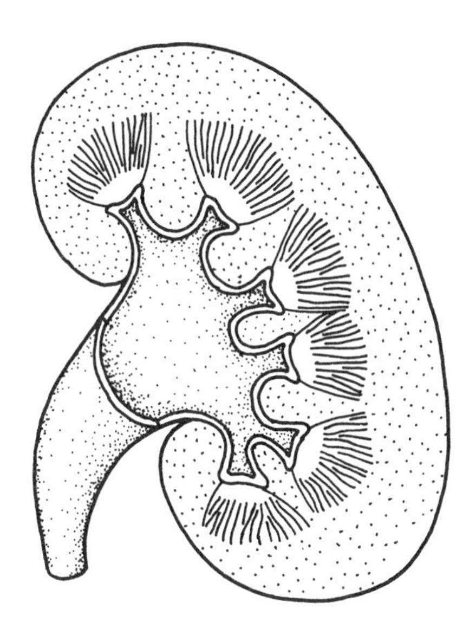

- - - - - - - - - - - - - - - - - - -

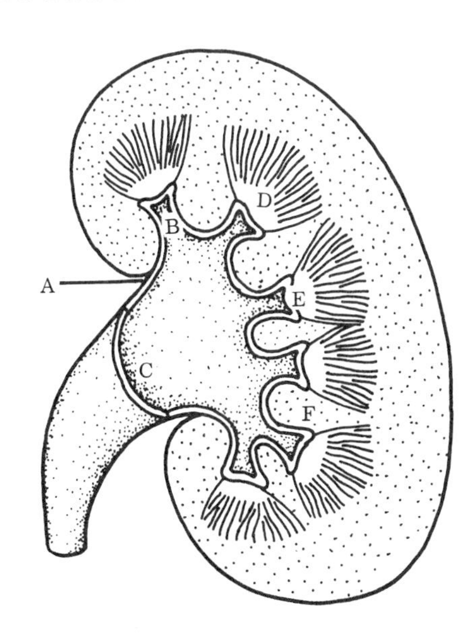

The nephron

6. The functional unit in the production of urine is a microscopic structure called the nephron. More than a million nephrons are located in each kidney.

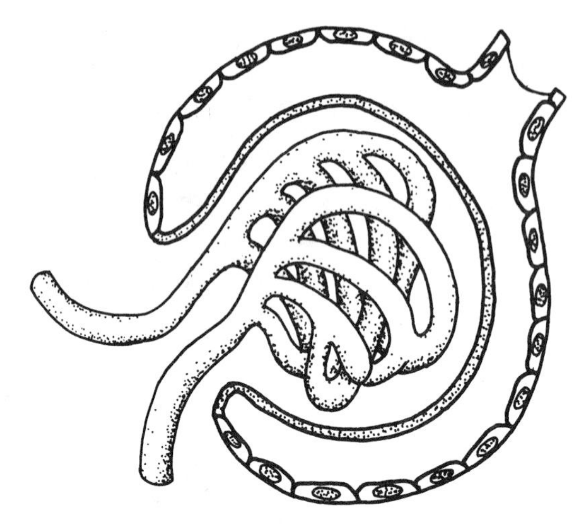

A nephron consists of a renal corpuscle, in which blood is filtered, and uriniferous (urine-making) tubules in which urine attains its final form. The blind end of a uriniferous tubule has a cuplike shape and surrounds a tuft of capillaries called the glomerulus. The cuplike capsule and the glomerulus make up the renal corpuscle; all renal corpuscles are located in the cortex. Materials are filtered from the blood in the capsule and passed into the tubule, which extends into the renal medulla.

(a) What is the functional unit of the kidney? ______________________

(b) What structures make up the renal corpuscle? ______________________

(c) In what structure are materials filtered from the blood? __________

____

(d) In what structure does urine attain its final form? ______________

(e) Would a glomerulus be found in a renal pyramid or in the renal cortex? ______________________

- - - - - - - - - - - - - - - - - - -

(a) nephron; (b) capsule and glomerulus; (c) corpuscle (from glomerulus into capsule); (d) uriniferous tubule; (e) renal cortex

7. Blood enters the kidneys through the renal arteries which branch directly from the aorta. The renal artery quickly divides into arterioles.

An afferent arteriole branches as it enters the corpuscle to produce each glomerulus. The capillary tuft then recombines to form an efferent arteriole leaving the glomerulus. These efferent arterioles form a capillary network surrounding the uriniferous tubule before forming the veins that join the renal vein. The small number of branches between the renal artery level and the glomerular capillary level results in a high blood pressure within the tuft. This facilitates the filtration process.

(a) Each afferent arteriole produces how many glomeruli? __________

(b) Name the vessel that leaves a glomerulus. ____________________

(c) In what two areas are capillaries found in a kidney? ____________

____

(d) What effect does increased pressure in renal arterioles have on the urine production? ________________________________

- - - - - - - - - - - - - - - - - -

(a) one; (b) efferent arteriole; (c) in each glomerulus; and in the region surrounding uriniferous tubule; (d) it increases filtration (Note: the glomerulus is the only capillary in the body with arterioles entering and leaving; most capillaries have arterioles leading in and venules leading out.)

8. The uriniferous tubule winds from the capsule of the renal corpuscle to empty eventually at a medullary papilla into a calyx. Just distal to the capsule is the proximal convoluted tubule. Here the tubule is curled around itself near the capsule. The epithelial lining is simple cuboidal with a microvillous border to increase the surface area for absorption.

The tube then forms a long straight U, the loop of Henle, with a distal convoluted tubule at the end. The uriniferous tubule then joins a collecting tubule to travel to the papilla. Capillaries closely follow and surround the area of the uriniferous tubule.

Along the length of the tubule, various components of the filtered material are resorbed back into the blood. The concentration differences between the materials inside the tubules and the capillaries is one cause of resorption. Hormones such as ADH (antidiuretic hormone) make the tubule more permeable to water. Thus more water is resorbed by the blood and a more concentrated urine results.

(a) Identify the structures indicated on the following diagram of a uriniferous tubule.

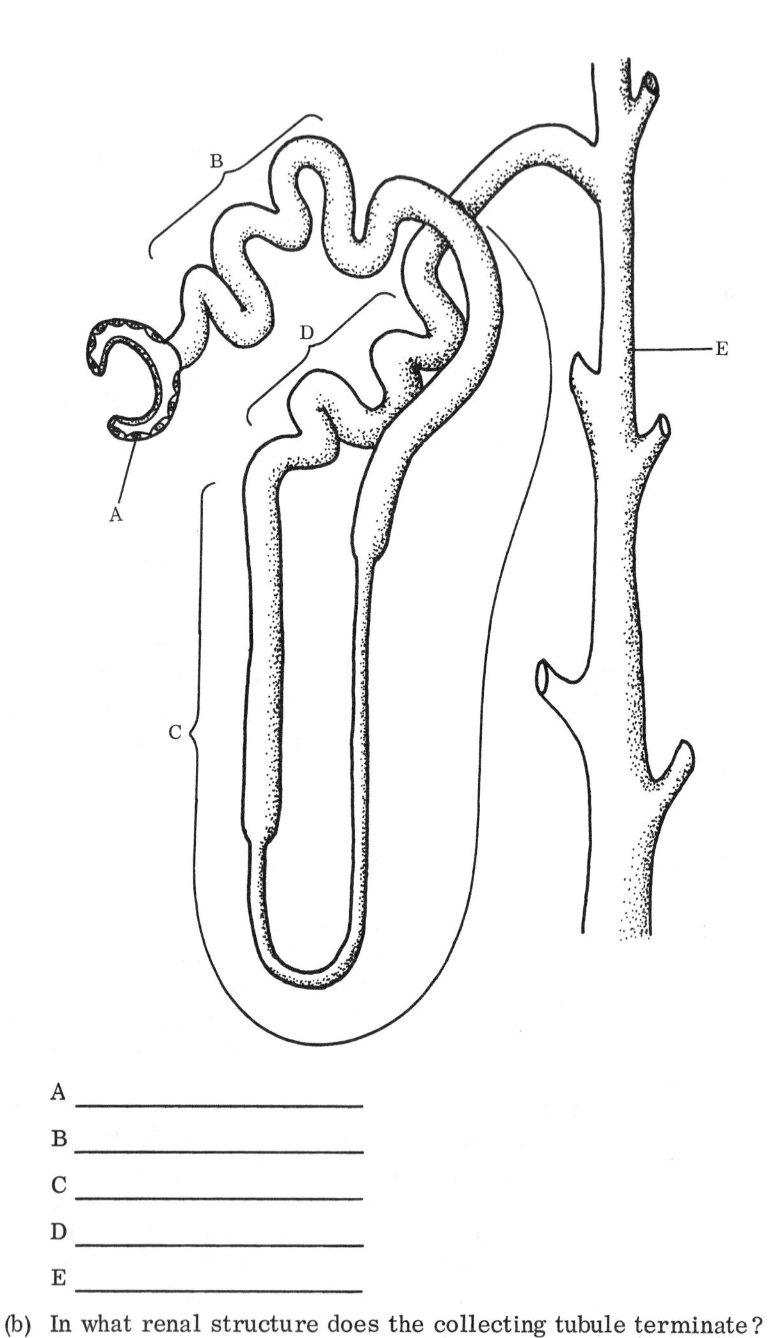

A ______________________

B ______________________

C ______________________

D ______________________

E ______________________

(b) In what renal structure does the collecting tubule terminate?

____

(c) Where are the capillaries located that resorb materials from urine?

____

(d) Is the ultimate effect of ADH to produce a dilute or concentrated urine? ____________________

- - - - - - - - - - - - - - - - - -

(a) A—capsule; B—proximal convoluted tubule; C—loop of Henle; D—distal convoluted tubule; E—collecting tubule
(b) papilla
(c) around loop of Henle and convoluted tubules
(d) concentrated

9. The microscopic features of the kidney correspond with some of its gross features. The cortex contains all of the glomeruli with capsules as well as the convoluted tubules. The medulla contains the loops of Henle as well as most of the collecting tubules. The renal columns contain arteries and veins that carry blood toward the glomerulus. In which area of the kidney (as indicated on the right) would you find the following?

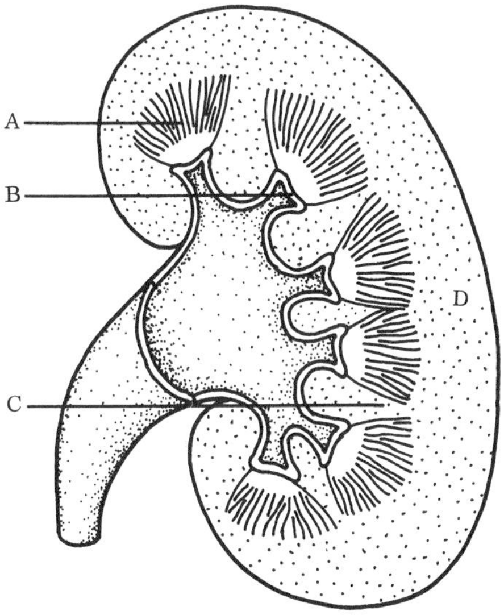

A
B
C
D

(a) glomeruli __________
(b) proximal convoluted tubules __________
(c) loops of Henle __________
(d) distal convoluted tubules __________
(e) the filtration mechanism __________
(f) the resorption mechanism __________
(g) the outlet from collecting tubules __________
(h) arteries and veins supplying the uriniferous mechanism __________

- - - - - - - - - - - - - - - - - -

(a) D; (b) D; (c) A; (d) D; (e) D; (f) A and D; (g) B; (h) C

URINE

10. As blood passes through a glomerulus, water and many plasma elements are filtered out into the double-walled capsule. The visceral layer of the capsule is closely adjacent to the capillary tuft to facilitate the filtration process. The parietal layer of the capsule is relatively impermeable, and contains the filtrate (provisional urine, or filtered blood elements), directing it into the uriniferous tubule. Selective reabsorption of most of the filtrate takes place along the tubule. The capillaries of the efferent arteriole enclose the tubule in a meshwork that aids in the reabsorption of nutrients, water, and other needed substances from the filtrate. The pressure and ionic differences between the blood and provisional urine play important roles in the reabsorption. Most of the water is resorbed from the proximal convoluted tubule, while NaCl (salt) balance is adjusted in the loop of Henle. The pituitary hormone ADH (antidiuretic hormone) acts by making the tubules more permeable to water, affecting the final quality of urine.

    (a) What are the two major processes in urine formation? ___________

    ____

    (b) Which portion of the tubule is most active in reabsorption of water?

    ____

    (c) What epithelial specialization is present in this part of the tubule?

    ____

    (d) Would increased secretion of ADH result in a more or less concentrated urine? ______________________

    (e) Are the capillaries that surround a uriniferous tubule branches of an efferent arteriole of an afferent arteriole? ______________

- - - - - - - - - - - - - - - - - -

(a) filtration and reabsorption; (b) proximal convoluted tubule; (c) microvilli (see frame 8); (d) more concentrated (more water is absorbed); (e) efferent

11. The concentrated urine is carried down a collecting tubule to a papilla, where the urine is emptied into a minor calyx of the pelvis. The minor calyces join to form four or five major calyces. These structures are found in the expanded upper end of a ureter.

    Each ureter is about 50 cm long, and has a wall composed of three layers. The mucosa is made up of transitional epithelium with its underlying lamina propria. A muscular coat has both an inner longitudinal layer and an outer circular layer of smooth muscle, which move the urine towards the bladder. An adventitia of connective tissue attaches the ureter to the posterior abdominal wall.

(a) What are the two components of the inner layer of the ureter wall?

____

(b) What are the two components of the muscular layer of the ureter wall? _______________

(c) What is the function of the muscular layer of the ureter? _______________

____

(d) What purpose does the adventitia of the ureter serve? _______________

____

- - - - - - - - - - - - - - - - - -

(a) transitional epithelium and lamina propria; (b) inner longitudinal and outer circular (this is opposite from digestive tract); (c) to move urine toward bladder; (d) to attach ureter to posterior wall of abdominal cavity

12. The ureters enter the bladder at two corners of a small area called the trigone; the urethral exit marks the third corner. The bladder changes in size substantially depending on the quantity of urine being stored. It changes rapidly at the time of urination, when 300 or 400 ml may be voided, leaving the bladder virtually empty.

Transitional epithelium lining the bladder allows for these changes. When the bladder is empty, seven or eight layers of cells are seen in its lining. These cells rearrange themselves as the bladder is distended, until only one or two layers are seen. The intercellular connections enable the epithelium, in effect, to make a transition between stratified and simple, depending on the needs. Three heavy layers of smooth muscle enable the bladder to expel the urine, controlled by a circular voluntary muscle at the urethral opening. Only the superior surface of the bladder is covered with peritoneum.

(a) How does the bladder's location affect its peritoneal covering?

____

(b) What openings mark the three corners of the trigone? _______________

____

(c) What feature of the bladder lining enables it to change its capacity?

____

- - - - - - - - - - - - - - - - - -

(a) the rest of the bladder is in the pelvic cavity, beneath the abdomen; (b) 2 ureters and 1 urethra; (c) transitional epithelium

13. The urethra conducts the urine from the bladder out of the body. It is of two forms, depending on the sex of the individual. In the female the urethra is about 4 cm long, while it is about 20 cm long in the male. As the male urethra will be examined in detail in the next chapter, only the female urethra will be considered here.

    The epithelial lining in the female urethra is transitional at the bladder end but is modified to stratified squamous at its junction with the skin. A circular muscular layer surrounds the epithelium and submucosa. The posterior part of the urethra is firmly attached to the anterior wall of the vagina; infection or injury to one of these organs is often transmitted to the other.

    (a) Name three organs involved in eliminating prepared urine from the body. ______________________________

    (b) What two distinct classifications of epithelium are present in the female urethra? ______________________________

    (c) To what pelvic organ (other than the bladder) is the female urethra firmly attached? ____________________

- - - - - - - - - - - - - - - - - -

(a) ureters, bladder, and urethra; (b) transitional and stratified squamous; (c) vagina

14. Let us consider some blood that is destined to be diverted from the aorta to contribute to urine formation.

    (a) Into what blood vessel from the aorta is it diverted? ____________

    (b) At what part of the kidney does this vessel enter? ____________

    (c) After several branches, the blood is in an afferent arteriole, ready to enter what structure? ____________________

    (d) What process takes place while the blood is at the location in (c)?

    ____

(e) After provisional urine is in the capsule, it passes into a tubule with three distinct areas. Name them. ____________________

____

(f) After the provisional urine is filtered out, what process takes place in the tubules? ____________________

(g) What part does the blood still in the vessels play in this? ________

____

(h) Urine in the collecting tubules is emptied into the ____________

____

- - - - - - - - - - - - - - - - - -

(a) renal artery
(b) hilum
(c) glomerulus
(d) filtration of water and proteins from blood
(e) proximal convoluted tubule, loop of Henle, and distal convoluted tubule
(f) reabsorption of water
(g) differences in concentrations between blood and provisional urine affect reabsorption
(h) minor calyx of pelvis of ureter

SELF-TEST

This Self-Test is designed to show how well you have mastered this chapter's objectives. Answer each question to the best ofyour ability. Correct answers and review instructions are given at the end of the test.

1. Identify the structures indicated in the figure at the right.

   A ____________________

   B ____________________

   C ____________________

   D ____________________

   E ____________________

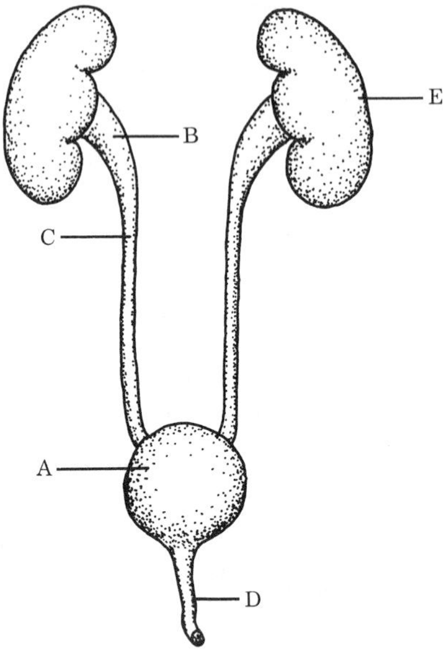

2. Name two structures found in the cortex of a kidney. ____________

   ____

3. What two processes are involved in production of urine? ____________

   ____

   ____

4. What makes up a renal corpuscle? ____________________

5. In what portion of a uriniferous tubule is most of the water reabsorbed?

   ____

6. What structures make up most of a medullary pyramid? ____________

   ____

7. What type of epithelium lines the following structures?

   (a) convoluted tubules ____________________

   (b) ureter and bladder ____________________

   (c) female urethra ____________________

Answers

Compare your answers to the Self-Test questions with those answers given below. If all of your answers are correct, you are ready to go on to the next chapter. If you missed any, review the frames indicated in parentheses following the answers. If you missed several questions, you should probably reread the entire chapter carefully.

1. A—bladder; B—pelvis; C—ureter; D—urethra; E—kidney (frames 2 and 3)
2. renal corpuscles (glomeruli) and convoluted tubules (frame 9)
3. filtration and reabsorption (frame 10)
4. capsule and glomerulus (frame 6)
5. proximal convoluted tubule (frame 10)
6. loop of Henle and collecting tubule (frame 9)
7. (a) cuboidal, microvilli (frame 8)
   (b) transitional (frame 11)
   (c) transitional and stratified squamous (frame 13)

PART V

Reproductive Systems of the Body

The systems you have studied so far enable the human body to live from day to day, to thrive as a separate organism. The remaining body system is the reproductive system, which continues the survival of the human race as a whole. The reproductive system differs in male and female individuals, but it has certain similarities in structure and function.

One cell from each type of reproductive system must unite to produce a new individual. You will see in your study of these last two chapters how these cells are produced, and how they travel to meet at the crucial time for conception.

The reproductive systems contribute more to an individual than just the reproductive capacity. Secondary sexual characteristics, such as voice, hair distribution, and fat distribution, are also regulated by products from the reproductive systems. Since certain reproductive organs also carry out an endocrine function, these two systems interact in the continuing existence of the human organism.

CHAPTER FOURTEEN
The Male Reproductive System

Just as maintenance of individual organisms is critical to their survival, so is maintenance of the species critical for survival of the human race. This last section deals with the male and female contributions to continuing the species. After you complete your study of this chapter, you will be able to:

- specify the location and function of the testes, epididymis, vas deferens, and ejaculatory duct;
- identify the location and contribution of the accessory glands—seminal vesicles, prostate, and bulbourethral (Cowper's) glands;
- specify the major structural components of the testes;
- differentiate between the microscopic structures of the testes, epidymis, and vas deferens;
- distinguish among the three portions of the male urethra;
- identify the source of fluids that comprise semen;
- trace the development of sperm;
- explain the mechanism of penile erection;
- name the source and the hormones that activate spermatogenesis and testosterone production.

The external structures of the male reproductive system, the penis and scrotum, are called the genitals; they are only a part of the entire system. They house the beginning and end of the system. Sperm develop in the testes, which are housed within the scrotum. They leave the male body via the urethra, found inside the penis. In this chapter we will cover the path from one end to the other.

THE TESTIS

1. The male reproductive system functions in producing sperm and transporting it to the outside. In addition, it produces a hormone called testosterone, that controls such secondary sexual characteristics as growth of facial and body hair, deepening of the voice, and distribution of fat. Sperm development and testosterone production both take place in the essential organ of male reproduction—the testis (plural, testes).

   A series of excretory ducts transport the sperm to the outside, while three types of accessory glands provide their secretions to aid in transportation, activation, or nourishment of the sperm. Supporting structures include penis and scrotum externally and the spermatic cord internally. The spermatic cord supports the duct system on its way into the abdominal wall.

   (a) Name two items produced in the testis. ____________________

   (b) Name one organ of the male reproductive system that belongs to each of the following categories.

   essential____________________

   supporting structure ____________________

   (c) What are effects of secretions of the accessory glands? __________

   ____

- - - - - - - - - - - - - - - - - -

(a) sperm and testosterone
(b) essential—testis, supporting structure—penis, scrotum, or spermatic cord
(c) aid in transporting, activating, or nourishing sperm

2. The following diagram shows the locations of some of the organs in the male reproductive system. You can see how they are related to the organs in the urinary system.

   The urethra functions in both systems. The testis is located within the scrotum and the epididymis caps the testis. The vas (or ductus) deferens then passes up from the testis into the pelvic cavity, and back to join the urethra with the duct of the seminal vesicle as the ejaculatory duct. The prostate gland empties its secretions at about the same level. Ducts from bulbourethral or Cowper's glands enter the urethra at a slightly lower level. The urethra then passes through the penis.

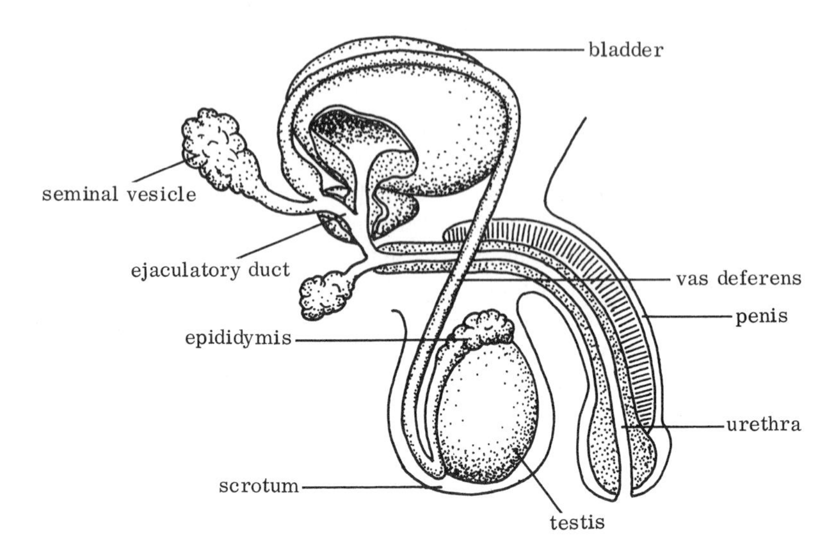

(a) The essential organ of the male reproductive system is located within what supporting structure?________________

(b) Name three segments of the excretory ducts of this system. ________

____

(c) List three accessory glands of this system. ________________

____

(d) Which of the three glands is closest to the distal end of the penis?

____

- - - - - - - - - - - - - - - - - - -

(a) scrotum
(b) vas deferens, ejaculatory duct, and urethra (epididymis, gland ducts)
(c) seminal vesicles, prostate, and bulbourethral glands
(d) bulbourethral

3. Since the testis is the essential male reproductive organ, it contains the origin of all sperm to be produced by the individual. Because each new life requires the union of a sperm from a male and an ovum from a female, the testicular contribution is vital. Before going into the structure of the testis, let us look briefly at the characteristics of a sperm —to learn just what makes it possible for a sperm to participate in the formation of a new individual.

In normal cell division, or mitosis, a cell produces a copy of its genetic material, and separates the genetic material. The cell then divides itself into two cells each with identical complete genetic material, in the form of 46 chromosomes.

In division of sex cells, however, the process of meiosis results in cells with half the previous amount of genetic material, or 23 chromosomes. The primary spermatocyte ("sperm cell") in the testis first undergoes normal mitosis into two secondary spermatocytes, each with a full amount of genetic material. Each secondary spermatocyte then undergoes meiosis, in which it separates its genetic supply in half, and divides without reproducing the genetic material. The result is four spermatids, each with the desired 23 chromosomes. These small nucleated structures then are gradually changed into spermatozoa (singular, spermatozoon, usually abbreviated as sperm) in the tubules of the testis.

(a) Add to the figure here the relevant number of cells and quantity of genetic material for each of the levels of spermatogenesis.

(b) In humans the normal number of chromosomes containing genetic material is 46. How many chromosomes are in a secondary spermatocyte? __________ In a spermatid? __________

(c) Two types of cell division occur in the production of sperm. Name the two types in the order in which they occur. ____________________

____

- - - - - - - - - - - - - - - - - -

(a)

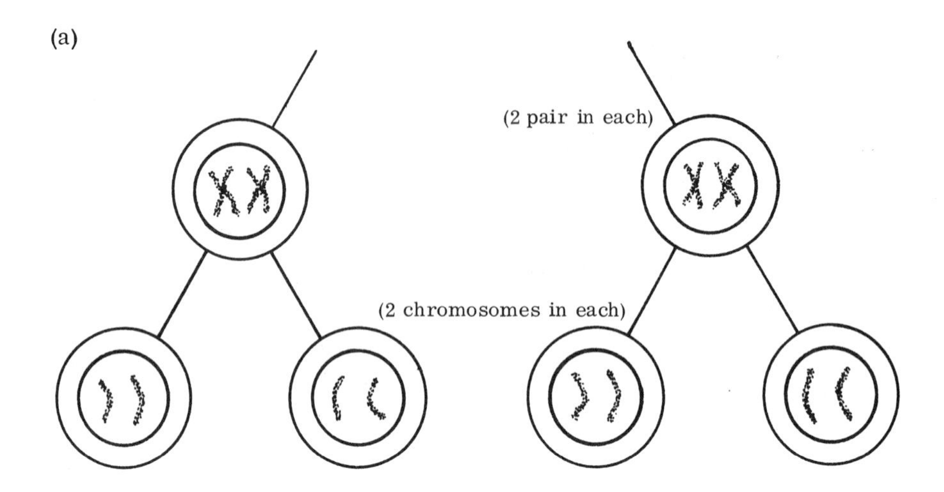

(b) 46; 23
(c) mitosis; meiosis

4. A mature spermatozoon is an elongated structure with a head and a relatively long tail. The ovoid head and long thin tail are joined by a middle piece, slightly thicker than the tail. The tail functions as a flagellum, or whip, propelling the sperm through fluids and eventually enabling one to reach an ovum. The spermatic head contains the half-supply of genetic material, while the tail provides the motility needed to join it with another half-supply.

   All the changes from primary spermatocyte to mature spermatozoon take place within the testis. About 250 tubules (the seminiferous tubules) make up each testis. After puberty, a section through a normal testis may show all of the stages of sperm development in the tubule wall. Immature cells (called spermatogonia) are pressed against the basement membrane. Next are a few layers of primary spermatocytes. Secondary spermatocytes are not seen very often because they undergo meiosis very quickly after they are formed. At the inner lining of the wall are spermatids, while mature spermatozoa are seen free in the central lumen, attached to supporting cells. They are ready to begin their journey through the tubule system.

   (a) Which of these drawings most resembles a spermatozoon? ________

   ____

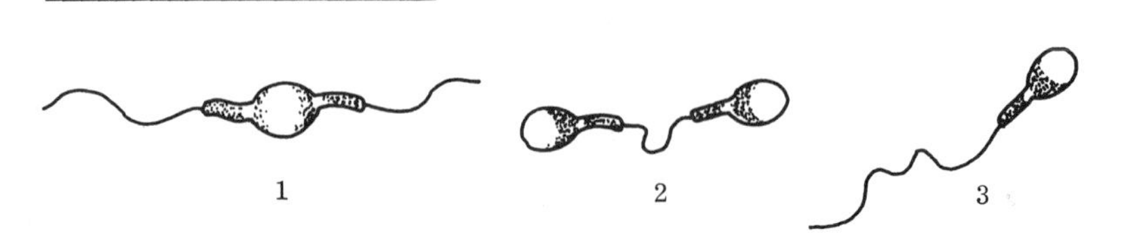

(b) In what part of a spermatozoon is the supply of genetic material located? ______________________

(c) The drawing at the right represents the wall of a tubule within the testis. Name the stages of sperm development found at the indicated levels.

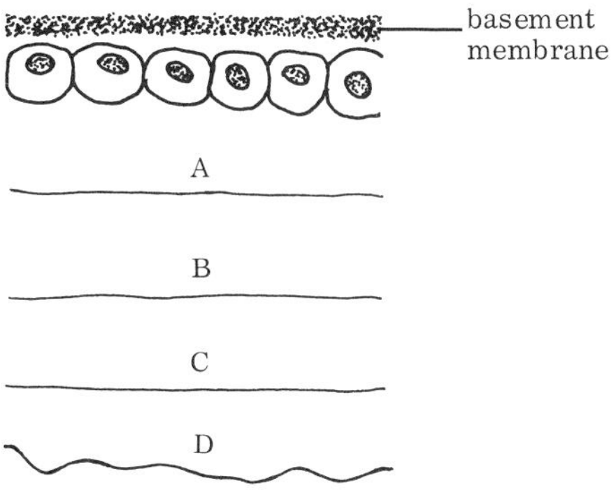

A ______________________

B ______________________

C ______________________

D ______________________

- - - - - - - - - - - - - - - - - -

(a) on the right
(b) head
(c) A—primary spermatocyte; B—secondary spermatocyte; C—spermatid; D—mature spermatozoon

5. Microscopically each testis is composed of about 250 segments called lobules, each containing one coiled seminiferous tubule. In addition to the germinal epithelium described in the previous frame, the tubule lining also includes many tall columnar cells (Sertoli cells) that nourish the developing sperm. In each lobule between the coils of the tubule are groups of interstitial cells that produce testosterone. These form the endocrine portion of the testis, controlling such features as secondary sex characteristics and sperm production. Connective tissue and blood vessels separate the lobules.

The collection of lobules that makes up a testis results in an ovoid mass, about 4-5 cm long by 2.5 cm wide. The covering adjacent to the lobules is a vascular coat from which the connective tissue that separates the lobules arises. External to the vascular coat is the tunica albuginea, forming a collagenous capsule attaching the testis to a portion of the abdominal lining that comprises its outer layer, the tunica vaginalis. The testis with its three-layered covering is further protected by the scrotum, a skin-covered muscular pouch divided into right and left compartments. The scrotal muscle is affected by temperature; it contracts when the external environment is cold to keep the developing sperm at the appropriate temperature. In hot surroundings, the muscle relaxes, to allow the testis to rest at a distance from the body heat.

(a) What is the function of tall columnar cells found in the seminiferous tubules? ______________________

(b) Where are the endocrine cells of the testis located? ______________

____

(c) What two structures make up the wall of the scrotum? ____________

____

(d) Name the three layers of the testis covering in order from the inside of the scrotum to the testis itself. ____________________________

____

\- - - - - - - - - - - - - - - - - -

(a) nourish developing sperm; (b) between coils of seminiferous tubules; (c) skin and muscle; (d) tunica vaginalis, tunica albuginea, and vascular coat

6. The seminiferous tubules from the testicular lobules join together into straight tubules which all lead to the same general area of the testis, the mediastinum testis. From this area, 12 to 16 efferent ductules carry spermatozoa into the epididymis for storage. Each epididymis consists of a single coiled duct, over 7 meters long if laid flat. The efferent ductules enter at the head of the epididymis.

   The epithelial lining is pseudostratified, and shows stereocilia, extensions that resemble a cross between cilia and microvilli. The stereocilia are thought to function in nourishing the free spermatozoa in the lumen of the epididymis (ductus epididymidis). A thin layer of smooth muscle surrounds the epithelium and may function in moving the spermatozoa to the tail of the epididymis, which is continuous with another duct, the vas deferens.

   (a) What purpose is served by the epididymis? ____________________

   (b) What is the function of stereocilia? ________________________

   (c) How does the epithelial lining of the epididymis differ from that of the testis? ____

   (d) What ducts carry spermatozoa into the head of the epididymis?

   ____

   (e) What duct carries spermatozoa from the tail of the epididymis?

   ____

\- - - - - - - - - - - - - - - - - -

(a) storage of spermatozoa
(b) nourish mature spermatozoa
(c) the epididymis epithelium is pseudostratified with stereocilia, while the testis has germinal epithelium with levels of sperm development
(d) efferent ductules
(e) vas deferens

CONDUCTING TUBES

7. The vas deferens (also called ductus deferens) passes upward from the tail of the epididymis along the posterior edge of the testis, through the inguinal canal and into the abdominal cavity. The inguinal canal is a slit in some abdominal muscles through which the testis descended before birth.

The vas deferens curves along the lateral surface of the bladder, turning inferiorly at its posterior edge. It then joins with the duct from a seminal vesicle (to be discussed next) to form the ejaculatory duct.

From the level of the testis to the inguinal canal, the course of the vas deferens is closely accompanied by nerves, blood vessels, lymphatic vessels, and cremaster muscle. All of these structures together make up the spermatic cord. The vas deferens' epithelium is similar to that of the epididymis; stereocilia gradually diminish, however, over its length.

(a) Name four structures that form part of the spermatic cord._______

____

(b) Indicate the order in which the excretory ducts pass the following structures.

__________ tail of epididymis

__________ duct of seminal vesicle

__________ inguinal canal

__________ testis

(c) How does the epithelium of the vas deferens differ at its own two ends? ____

- - - - - - - - - - - - - - - - - - -

(a) vas deferens, nerves, blood vessels, lymph vessels, and cremaster muscle (any four)
(b) 1, 4, 3, 2
(c) fewer stereocilia at distal end (near ejaculatory duct)

8. The seminal vesicles are located one on each side behind and beneath the bladder. These glands secrete a substance that nourishes the sperm and provides most of the fluid in which sperm are ejaculated. The duct from the seminal vesicle joins with the vas deferens to become the ejaculatory duct. The two ejaculatory ducts enter the posterior surface of the prostate gland and pass through it before opening into the urethra.

   The prostate is a doughnut-shaped gland that surrounds the urethra as it leaves the bladder. Its secretions enter the urethra through as many as 30 ducts and are thought to activate the sperm, which have been immobile up to now.

   The alkaline secretions of the bulbourethral (Cowper's) glands enter the urethra through separated ducts. These secretions may counteract the effect of acid urine, maintaining a satisfactory environment for the now-active spermatozoa.

   (a) Give an effect of the secretion of each of the following glands.

   seminal vesicle ____________________

   prostate ____________________

   bulbourethral gland ____________________

   (b) What gland empties into the ejaculatory duct? ____________________

   What else empties into this duct? ____________________

   (c) In the diagram at the right, label the accessory glands of the male reproductive system.

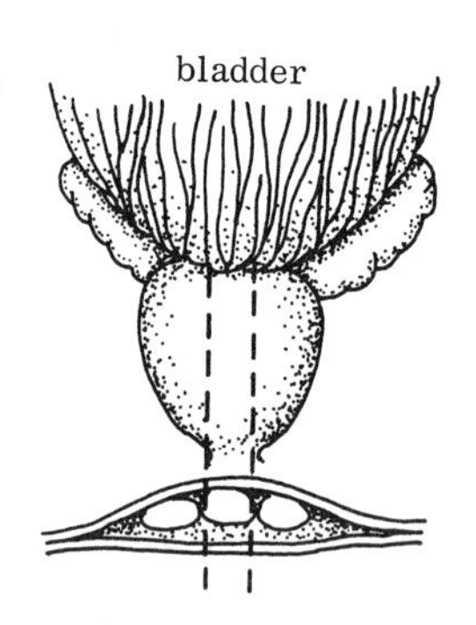

- - - - - - - - - - - - - - - -

(a) maintain potential for motility; activate spermatozoa; neutralize acid in urethra
(b) seminal vesicle; vas deferens
(c) (See figure at right.)

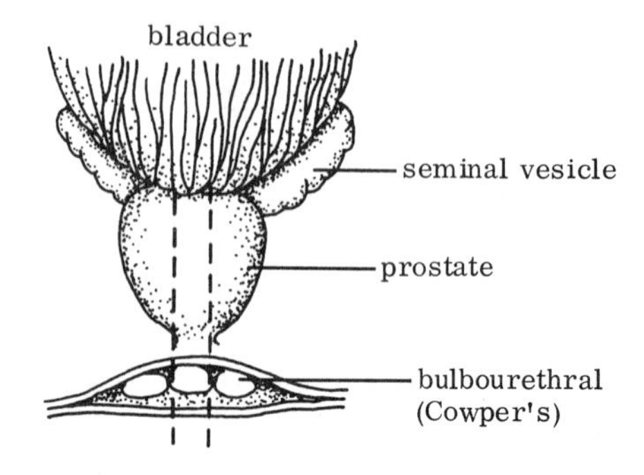

9. Ejaculated semen contains about 100 million spermatozoa in each cubic centimeter. Secretions from the epididymis, seminal vesicles, prostate, and bulbourethral glands suspend the spermatozoa, activating it and maintaining an appropriate environment. The bulk of semen, about 60 percent, is made up of seminal vesicle secretions.

Ejaculation results from contraction of the smooth muscle covering of the prostate gland and of the bulbocavernous muscle which compresses part of the urethra. Peristalsis in the bulbourethral and ejaculatory ducts also contributes to the process. A normal ejaculation consists of about 2 or 3 ml of semen, containing over 200 million sperm.

The fertility of the male depends not only on the number of sperm, but also on the activity level. It takes only one sperm to fertilize an ovum, but that sperm must travel quickly for it loses its power in about 24 hours. They actually move at a rate of 1.5 mm per minute.

(a) Which accessory gland contributes secretions that make up most of the semen? ______________________

(b) Name three anatomical structures that contribute to the process of ejaculation? ______________________

(c) What quantity of semen is produced in a typical ejaculation? ______________________ How many sperm would it contain (roughly)? ______________________

- - - - - - - - - - - - - - - - - -

(a) seminal vesicle
(b) bulbocavernous muscle, muscle of prostate covering, ejaculatory ducts peristalsis, bulbourethral duct peristalsis (any three)
(c) 2-3 ml; 200-300 million (or more than 200 million)

THE URETHRA

10. The male urethra is the terminal organ for both the urinary and reproductive systems. It extends from the outlet in the bladder to the distal end of the penis, and has three portions.

The prostatic urethra is directly below the bladder, and it is surrounded by the prostate gland. The ducts of the prostate, as well as the right and left ejaculatory ducts, open into the prostatic urethra. It is lined with transitional epithelium as is the adjacent bladder. The membranous portion of the urethra is very short, just 1-2 cm long, and passes through the pelvic floor. The bulbourethral gland ducts open into it. Psuedostratified epithelium lines the membranous urethra.

The third portion, the penile (cavernous) urethra is usually 15 cm long, and is entirely within the penis. It is also lined with pseudostratified epithelium. Mucous cells are interspersed among the pseudostratified epithelium.

(a) If the prostatic urethra is 3 cm long, what is the length of the male urethra? ____________

(b) Into which segment of the urethra does the seminal vesicle empty? ____________

(c) What type of epithelium lines each of the portions of the urethra? ____________

(d) What ducts empty into the membraneous urethra? ____________ Through what structure does it pass? ____________

- - - - - - - - - - - - - - - - - - - -

(a) 19–20 cm
(b) prostatic
(c) prostatic—transitional; membranous—pseudostratified; penile (cavernous)—pseudostratified
(d) bulbourethral gland ducts; passes through pelvic floor

11. The penis functions as a reproductive organ in enabling semen to be delivered to the site of choice. Its structure is formed from three cavernous bodies containing many openings that can fill with blood. Two posterior bodies are called the corpora cavernosa (singular, corpus cavernosum), while the anterior body is called the corpus spongiosum. Actually, all are similar in structure, but the cavernous bodies are larger while corpus spongiosum houses the urethra. The bulbous head of the penis, called glans penis, is an extension of corpus spongiosum.

The three bodies are supported and held together by connective tissue and skin. The skin at the end of the penis is folded back on itself to cover the glands. This is called the foreskin or prepuce and is often removed from newborn male babies in a circumcision.

Blood vessels to the penis are muscular arteries that empty into the sinuses of the three cavernous bodies. As they fill, the veins that drain the penis are compressed. Blood cannot escape, yet more blood pours in. The penis becomes erect and turgid. With physical relaxation, the blood will flow out again, usually after ejaculation.

(a) Name the three bodies that make up the bulk of the penis. ____________
____

(b) What term describes the bulbous structure at the distal end of the penis? ____________________

(c) Which body of the penis houses the cavernous urethra? __________ ____________________

(d) What part of the penis is removed in a circumcision? __________ ____________________

(e) Describe the physical process that causes the penis to become erect. ____________________

- - - - - - - - - - - - - - - - - - -

(a) two corpora cavernosa and one corpus spongiosum
(b) glans penis
(c) corpus spongiosum
(d) prepuce (foreskin)
(e) Blood from arteries pours into spaces in the corpora of the penis. This compresses the veins, blocking outflow of blood. Blood continues to flow in, causing the penis to become full and erect.

12. As we discussed in Chapter 7, the testes have an endocrine function. The hormone testosterone is secreted by the interstitial cells between the coiled tubules in testicular lobules. Testosterone controls the development and maintenance of secondary sex characteristics in the male —hair distribution, voice changes, and muscular development, as well as adolescent maturing of ducts and glands of the reproductive system.

Testosterone production is, in turn, controlled by a hormone of the anterior pituitary gland, ICSH, which stimulates the interstitial cells to secrete testosterone. The level of testosterone in the blood at a given time exerts a negative feedback effect on ICSH production; a high testosterone level results in decreased production of ICSH. Another hormone of the anterior pituitary is follicle-stimulating hormone (FSH); FSH controls spermatogenesis, or the development of spermatozoa.

(a) What endocrine gland controls the hormonal action of the testes? ____________________

(b) A man whose secondary sexual characteristics are not developed would probably have a deficiency of what pituitary hormone? ______ ____________________

(c) A low level of testosterone in the blood would have what effect on the anterior pituitary of a normal male? ____________________

(d) What endocrine hormone controls spermatogenesis? ____________

____

- - - - - - - - - - - - - - - - - - -

(a) anterior pituitary (adenohypophysis); (b) ICSH; (c) stimulate it to increase production of ICSH; (d) FSH

Undescended testicles

During embryonic development of a male, the testes are located within the abdominal cavity. About two months before birth, they usually descend into the scrotum through the inguinal canal. Occasionally, the testes remain in the abdomen. If not corrected surgically, this condition, called cryptorchism, causes sterility since living sperm cannot be produced at internal body temperature. Cryptorchism does not affect the production of testosterone. When the testes are brought down into the scrotum surgically before puberty, the male normally will be fertile.

Vasectomy

As a method of permanent birth control for men, the vasectomy is gaining in popularity. A small incision is made above each spermatic cord. The vas deferens is cut, and the two ends are tied off. No spermatozoa are then able to reach the penis for ejaculation. The glandular contribution that makes up the bulk of the semen remains the same.

Enlarged prostate

An enlarged prostate gland commonly causes men of advancing years to experience difficulty in urinating as the urethra is compressed. Whether the cause is benign (prostatic hypertrophy) or malignant, surgery to remove at least the medial portions of the gland is required.

SELF-TEST

This Self-Test is designed to show how well you have mastered this chapter's objectives. Answer each question to the best of your ability. Correct answers and review instructions are given at the end of the test.

1. Indicate the sequence in which sperm pass through the following structures.

   __________testis

   __________urethra

   __________ejaculatory duct

   __________epididymis

   __________vas deferens

2. How does the wall of a seminiferous tubule differ from the epididymal wall? ______________________________

3. Name the longest segment of the urethra. ______________________________

   What accessory gland(s) open into it? ______________________________

4. What is the source of the largest component of semen? ______________________________

5. What are three developmental stages of spermatozoa? ______________________________

6. Which endocrine gland controls the endocrine function of the testis? ______________________________

7. At what location (approximate) in the male reproductive tract do sperm become active? ______________________________

8. Where are stereocilia found? ______________________________

   What is their function? ______________________________

9. How does the structure of the penis allow for erection? ______________________________

10. How do you think a vasectomy affects sperm production? ______________________________

Answers

Compare your answers to the Self-Test questions with those answers given below. If all of your answers are correct, you are ready to go on to the next chapter. If you missed any, review the frames indicated in parentheses following the answers. If you missed several questions, you should probably reread the entire chapter carefully.

1. 1, 5, 4, 2, 3 (frame 2)
2. germinal epithelium in seminiferous tubule compared to pseudostratified in epididymis (frames 5 and 6)
3. cavernous urethra; none (frame 10)
4. prostate gland (frame 8)
5. primary spermatocyte, secondary spermatocyte, and spermatid; before spermatid (frame 3)
6. pituitary (anterior, or adenohypophysis) (frame 12)
7. prostatic urethra (frame 8)
8. epididymis; nourish sperm (frame 6)
9. sinuses fill with blood, compressing the veins (frame 11)
10. It doesn't; it prevents their transport to the penis. (frame 7)

CHAPTER FIFTEEN

The Female Reproductive System

The reproductive system in the female is both similar to and different from that of the male. Both systems produce a germ cell with half the typical chromosomal number. Both include paired endocrine glands under the control of pituitary hormones. Both have a system of tubes or passageways for movement of the germ cell. In spite of these similarities, however, the differences are very striking, and shall be the focus of this chapter. After you complete your study of this chapter, you will be able to:

- identify the organs of the female reproductive system and give their functions;
- describe the layers of the uterine wall;
- locate the female reproductive organs in relation to the urinary and digestive excretory organs;
- state the sources of estrogen and progesterone;
- describe the stages in the maturation of ova;
- correlate the ovarian and endometrial cycles;
- relate the endometrium to the phases in the menstrual cycle;
- specify normal times for fertilization and implantation.

THE REPRODUCTIVE ORGANS

1. The female reproductive system has both internal and external components. On the following page is a diagram of the external components, or genitalia.

   The major and minor lips (or labia) protect the vaginal and urethral openings. These are folds of mucous membrane and skin. The anterior of the minor lips join at the clitoris, which is a sensory organ. The entire area from the hair-covered mons veneris to the anus is called the vulva.

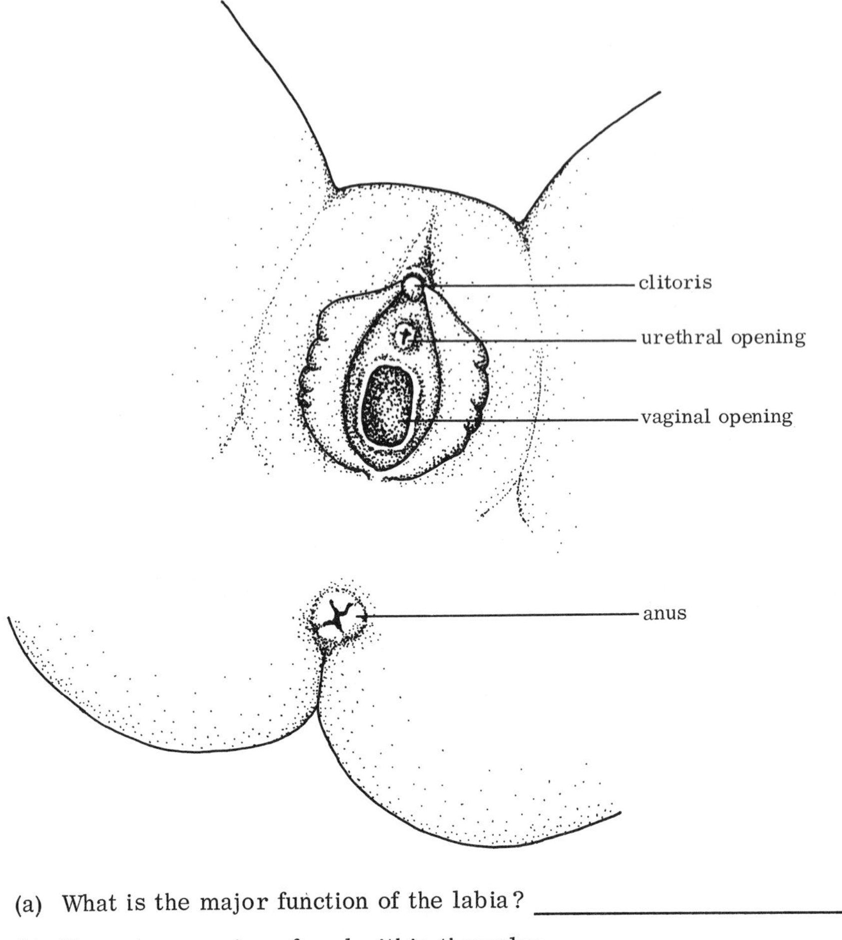

(a) What is the major function of the labia? ____________________

(b) Name two openings found within the vulva. ____________________

(c) Which set of lips are more lateral in position? ____________________

- - - - - - - - - - - - - - - - - - -

(a) protection; (b) urethral and vaginal; (c) major lips

2. The following diagram shows the internal organs of the female reproductive system. The paired ovaries are endocrine glands that secrete the female sex hormones. In addition, meiosis and maturation of the sex cell, or ovum, takes place in the ovary. The uterine tubes (Fallopian tubes) extend from near the ovary to the uterus itself. The uterus is pear-shaped, with its apex at the cervix or neck. The cervix extends about an inch into the vagina, which opens to the surface between the minor lips.

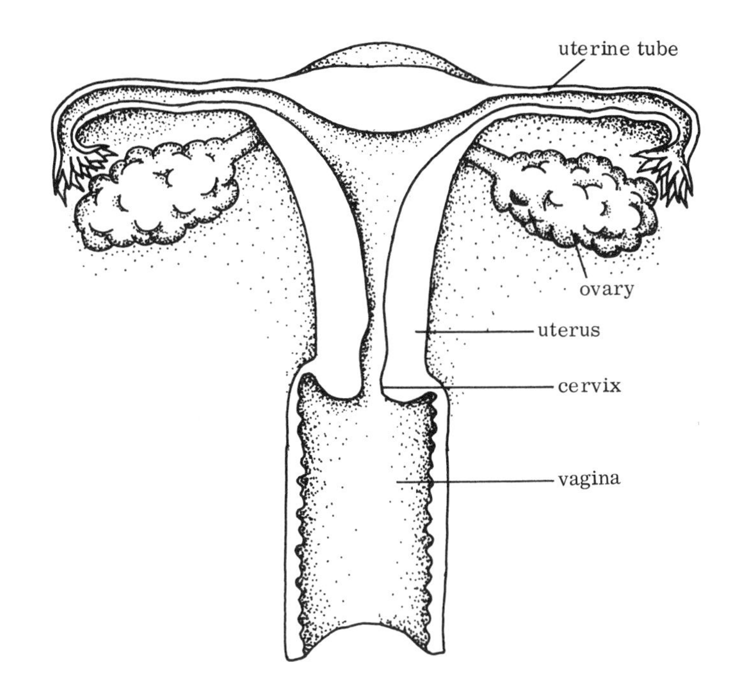

(a) In which organ are ova produced? ______________________

(b) Name the organ that enables ova to travel from an ovary to the uterus. ______________________

(c) The cervix is a part of which organ? ______________________

(d) Which internal female reproductive organ extends to the exterior?

____

- - - - - - - - - - - - - - - - - - -

(a) ovary; (b) uterine (Fallopian) tube; (c) uterus; (d) vagina

3. The vagina connects the external female genitalia to the uterus. It extends from between the minor lips to the cervix of the uterus, existing as a potential space with deep folds in the mucous membrane lining. The vagina is 10 to 15 cm long, and is located in the midline, between the bladder and rectum. Its wall includes a muscular layer capable of extreme dilation in addition to the deeply folded mucous lining. The vagina has two functions. It serves as the female organ for sexual intercourse, and it serves as the last portion of the birth canal.

(a) Label the bladder, vagina, and rectum in the diagram at the right.

(b) In which of its functions does the vagina dilate or stretch the most?

____

(c) What is the normal vaginal length? ______________

(d) What is the width of the vagina at rest? __________

____

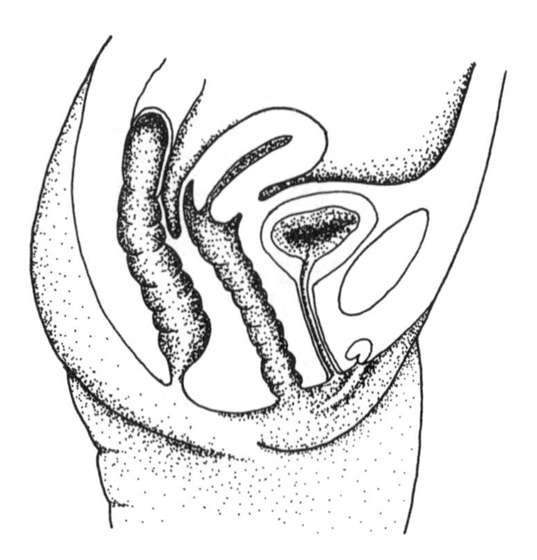

- - - - - - - - - - - - - - - - - -

(a)

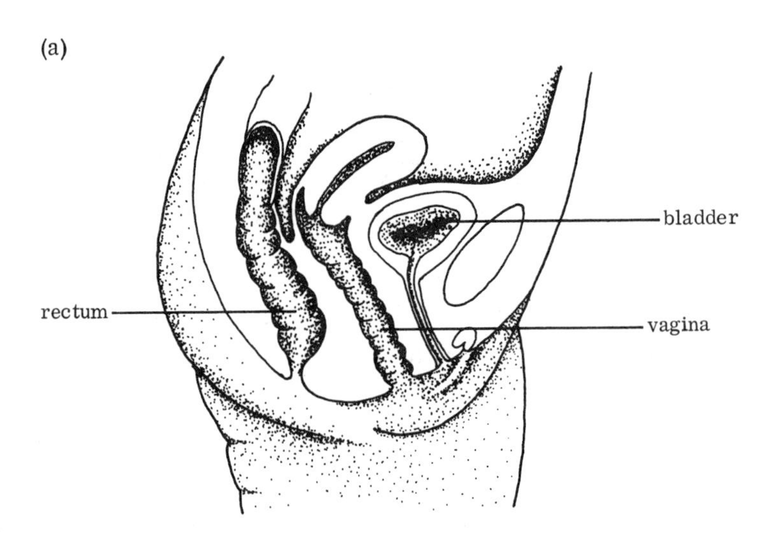

(b) birth canal
(c) 10–15 cm
(d) minimal—the sides rest against each other

4. The uterus is shaped like an upside-down pear whose apex, the cervix, projects into the vagina. Secretions of the projecting cervix keep the vagina moist. The body of the uterus is above the cervix and tilts forward above the bladder. The right and left sides of the uterine body are continuous with the uterine tubes. The uterine cavity is triangular, with points at the two tubes and at the cervix. It is normally flattened in an anterior-posterior direction.

The uterus is supported in the pelvic cavity by several ligaments. Of these, the broad ligaments are most significant; they fix the uterus to both sides of the pelvic cavity and contain the uterine arteries that supply the organ.

(a) Which part of the uterus produces the secretions that moisten the vagina? ____________________

(b) Name the arteries that supply the uterus. ____________________

(c) List the organs at each of the three openings into the uterus. ______

____

(d) Where is the body of the uterus located with respect to the bladder?

____

- - - - - - - - - - - - - - - - - -

(a) cervix; (b) uterine; (c) right uterine tube, left uterine tube, and vagina; (d) above the bladder (tilting forward)

5. The wall of the uterus has three major components. The thin outer layer is an extension of the peritoneum of the abdominal cavity. The thickest layer is in the middle; the myometrium is made up of several layers of intertwining smooth muscle fibers. The inner layer, the endometrium, consists of an epithelial lining with mucous cells and underlying connective tissue. This layer is highly vascular, and varies in thickness and appearance according to the stage in the menstrual cycle. (The cycle will be covered in detail later.)

(a) What type of tissue forms the inner lining of the uterus? __________

____

(b) Would you describe the outer layer of the uterine wall as fibrous, serous, or muscular? ____________________

(c) Name the layer of the uterine wall that shows variation at different points in the menstrual cycle. ____________________

(d) Name the thickest layer of the uterine wall. ____________________

(e) What type of tissue makes up this thickest layer? ______________

- - - - - - - - - - - - - - - - - -

(a) epithelial; (b) serous (extension of peritoneum); (c) endometrium; (d) myometrium; (e) smooth muscle

6. The uterine tubes extend from the body of the uterus toward the lateral wall of the pelvis. Each is about 10 cm long and wider at the free end which opens near the ovary. Many fingerlike projections, called fimbriae, surround the opening of the tube into the abdominal cavity.

   The three layers of the tube wall are continuous with the walls of the uterus. The inner layer, however, is different. The tubal lining does not reflect the menstrual cycle. And cilia are present throughout most of the tube. The fimbriae on the free end aid in drawing an ovum into the tube; the cilia work with muscular peristalsis to move it toward the uterus.

   (a) Which layer of the wall of the uterine tube is not involved in movement of ova? ____________________

   (b) How do the cilia of a uterine tube differ in function from the stereocilia of the tubes in the male reproductive system? ____________

   ____

   (c) What portion of a uterine tube would you expect to be physically located closest to an ovary? ______________________________

- - - - - - - - - - - - - - - - - -

(a) serous (cilia and muscle both help); (b) cilia aid movement, while stereocilia are nutritive; (b) fimbriae of distal end

7. The ovaries are the heart of the female reproductive system, just as the testes are in the male. An ovary is ovoid, 3 to 4 cm long, located in the upper pelvis near the end of a uterine tube.

   The ovary is covered by a layer of cuboidal epithelium. Just beneath the epithelium is a connective tissue layer, tunica albuginea, which corresponds to this layer in the male. Beneath this covering the ovary has a medulla made up of connective tissue, blood vessels, and supportive cells, with the cortex in the center containing follicles at various stages of development. Each follicle is a site of ovum maturation. In addition to ova development, the ovary is responsible for production of sex hormones, estrogen, and progesterone.

   (a) In which part of the ovary are follicles found? ________________

   (b) What type of tissue makes up the ovarian tunica albuginea? _______

   ____

   (c) What gland controls the secretion of the ovarian hormones? _____

   ____

- -

(a) cortex; (b) connective tissue; (c) pituitary (anterior, or adeno-hypophysis)

OVUM PRODUCTION

8. In a developing female embryo, as many as 400,000 primary ova may sink from the germinal epithelium, where they originate, into the cortex of the ovary, where they may later develop into functional ova. Since a woman ovulates only about 400 times during her fertile years, the back-up supply is tremendous. Many of the ova degenerate, or undergo atresia, even before puberty, but most become atretic later in life.

   The primordial follicles consist of a single layer of flattened epithelial cells surrounding each oocyte. (This corresponds to the spermatocyte.) Each oocyte has a large nucleus with a distinct nucleolus. As a follicle begins to mature, a primary follicle is formed as the flattened cells proliferate until stratified cuboidal epithelium encases the oocyte. When a cavity or antrum is formed within the epithelium and this antrum becomes filled with fluid, a secondary follicle has been formed.

   The connective tissue around the follicle also undergoes changes, and begins to function as an endocrine gland. The ovum continues to increase in size as these changes take place, reaching its largest size as the antrum is completed.

   (a) How can you distinguish between a primordial, a primary, and a secondary follicle? ______________________________________

   ____

   (b) What tissue in the ovary performs the endocrine function—the ovum, the epithelium, or the connective tissue? ____________________

   (c) Where do the oocytes in the ovarian cortex come from? __________

   ____

- - - - - - - - - - - - - - - - - - -

(a) secondary has antrum, primary does not; primordial is surrounded by one layer of flattened (squamous) cells, primary by one or more layers of cuboidal cells, and secondary by stratified cuboidal
(b) connective tissue
(c) germinal layer

9. Shortly before it is released, the oocyte begins to divide to attain the half-amount of genetic material needed for reproduction. Meiosis occurs first, resulting in two cells with equal number of chromosomes, but

very unequal cytoplasm. The larger secondary oocyte has most of the cytoplasm, while the smaller polar body is basically a package of genetic material.

At the time of release of the ovum, a maturation division occurs, producing another polar body. The first polar body may also divide at this time. All of the polar bodies degenerate and thus the original primary oocyte results in one functional ovum.

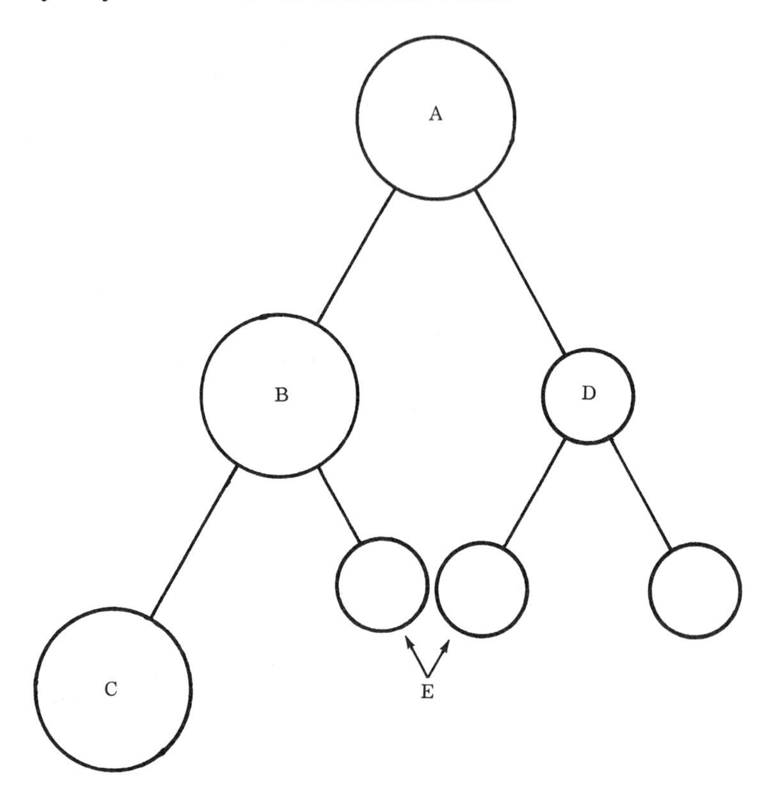

(a) Label the indicated components in the diagram above.

A ____________________

B ____________________

C ____________________

D ____________________

E ____________________

(b) What is the basic difference between a secondary oocyte and the first polar body? ____

____

(c) How does the ovumproduction process differ from the spermatozoon production process in the number of functional cells produced?

____

(d) How many chromosomes are present in a primary oocyte? ______

In a released ovum? ________

- - - - - - - - - - - - - - - - - -

(a) A—primary oocyte; B—secondary oocyte; C—ovum; D—polar body; E—polar body
(b) oocyte has more cytoplasm
(c) spermatocyte produces 4 sperm, while oocyte produces 1 ovum
(d) 46, 23

THE OVARIAN CYCLE

10. At approximately 28-day intervals, a mature follicle ruptures, at the time of the second division. In this process, called ovulation, the ovum is released into the abdominal cavity. The ruptured follicle then collapses into the ovary, and becomes a temporary gland, the corpus luteum (yellow body), which secretes progesterone.

Usually, the corpus luteum begins to degenerate after about a week, and is replaced by scar tissue (a form of connective tissue) to be called the corpus albicans (white body). If pregnancy ensues, the corpus luteum continues its glandular function, secreting estrogen in addition to progesterone, until the pregnancy is over.

(a) At ovulation, which structure forms the corpus luteum? ________

____

(b) Of what type of tissue is the corpus albicans formed? __________

____

(c) What is the primary hormone secreted by the corpus luteum? ____

____

(d) What is the duration of a corpus luteum under normal (non-pregnant) conditions? ________________

- - - - - - - - - - - - - - - - - -

(a) collapsed follicle; (b) connective tissue (scar tissue); (c) progesterone; (d) about one week

11. When an ovum is released into the abdominal cavity, it is normally entrapped by the fimbriae of the adjacent uterine tube. Ciliary action and peristalsis move the ovum inexorably toward the uterus; spermatozoa, if present, move upstream. The ovum must be joined by a sperm within 24 hours for pregnancy to occur; this fertilization process may take place anywhere in the tube. If fertilization does take place, three more days pass before the structure reaches the uterus. By about a week after ovulation, a fertilized ovum will implant in the uterine wall. If the ovum is not fertilized it normally passes out of the body unnoticed.

    (a) Name three structures active in moving the ovum from the ovary to the uterus. ______________________

    (b) A fertilized ovum is implanted in the uterus about seven days after ovulation. What occurs in the ovary at this time when the ovum has not been fertilized? ______________________

    (c) In what portion of the female reproductive system does fertilization normally occur? ______________________

- - - - - - - - - - - - - - - - - -

(a) fimbriae, cilia, and smooth muscle; (b) corpus luteum begins to degenerate; (c) uterine tube

12. Under non-pregnant conditions, the ovarian hormones exhibit a regular cycle. The cycle begins with an increase in FSH secretion in the pituitary, which stimulates a follicle to mature. As the antrum is formed, LH (luteotrophic hormone) secretion in the pituitary increases. LH works with FSH, and is critical for maturation and ovulation. A final spurt of LH immediately precedes ovulation, and initiates the glandular activity of the collapsed follicle. The corpus luteum secretes a small amount of estrogen; the growing follicle had secreted more estrogen, which inhibited FSH while stimulating LH production. The corpus luteum secretions of progesterone then inhibit LH production. As the corpus luteum degenerates, estrogen and progesterone levels decrease; the resultant increase in FSH begins another cycle. (See the diagram of this hormone cycle on the following page.)

    (a) Which pituitary hormone results in increased estrogen levels?

    ____

    (b) What is the immediate cause of corpus luteum degeneration?

    ____

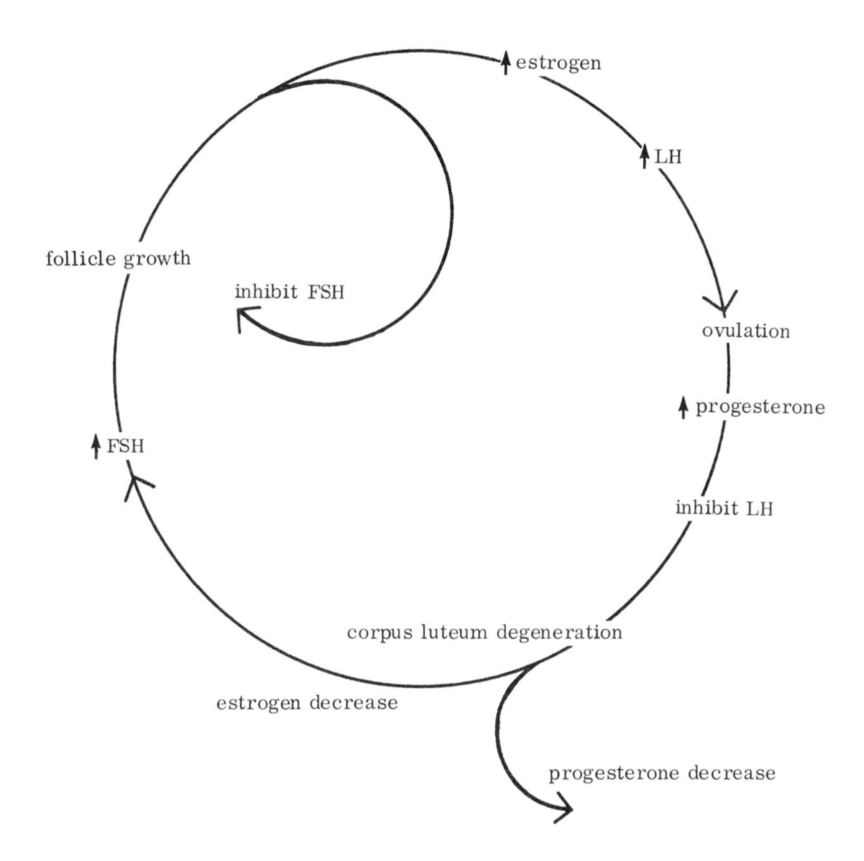

(c) If fertilization takes place, the above cycle might be altered before which point? ____________________

- - - - - - - - - - - - - - - - - -

(a) FSH, LH a bit; (b) decrease in LH levels; (c) corpus luteum degeneration

THE UTERINE CYCLE

13. Concurrently with the ovarian cycle, a menstrual cycle takes place in the non-pregnant uterus. This cycle is of the same length, as the two are intimately interrelated.

    The menstrual cycle is usually considered as beginning the first day of menstruation. The menstrual phase lasts three to five days. During this phase most of the epithelial lining with blood vessels and glands is shed. The estrogenic (proliferative) phase follows, in which the lining is rebuilt for ten days. The progesteronal (secretory) phase lasts for another ten days. In this phase, the epithelial glands produce glycogen and mucin, ready to nourish an implanted fertilized ovum, should one appear. When pregnancy does not occur, the two or three day premenstrual phase results. Decreased hormones cause vascular changes, and constricted arteries quickly lead into the menstrual phase once more.

(a) The ovarian cycle began with increased FSH production. What phase of the menstrual cycle might this correlate with? ________________

(b) During what phase of the menstrual cycle would you expect ovulation to occur? ________________________

(c) What ovarian cycle incident causes the premenstrual phase to begin?

____

(d) A fertilized ovum would become implanted during which phase of the menstrual cycle? ______________________

(e) What takes place in the endometrium during the estrogenic phase?

____

- - - - - - - - - - - - - - - - - -

(a) menstrual; (b) estrogenic; (c) degeneration of corpus luteum; (d) progestational; (e) build up of tissues, including arteries and glands

14. The chart on the following page summarizes the ovarian and menstrual cyles, showing how they are interrelated.

The cycles continue uninterrupted until fertilization takes place. When a sperm and ovum unite, a single cell with normal genetic material (46 chromosomes) is produced. This begins dividing immediately, and is a sphere of cells by the time it implants itself during the progestational phase.

The implantation has an effect on the corpus luteum of the follicle in which that ovum matured. Instead of degenerating immediately, it continues to secrete the hormones for maintenance of pregnancy. The secretion is highest during the first three months, but continues at a reduced level throughout the pregnancy. After that, the corpus luteum is converted to a corpus albicans as its function is complete.

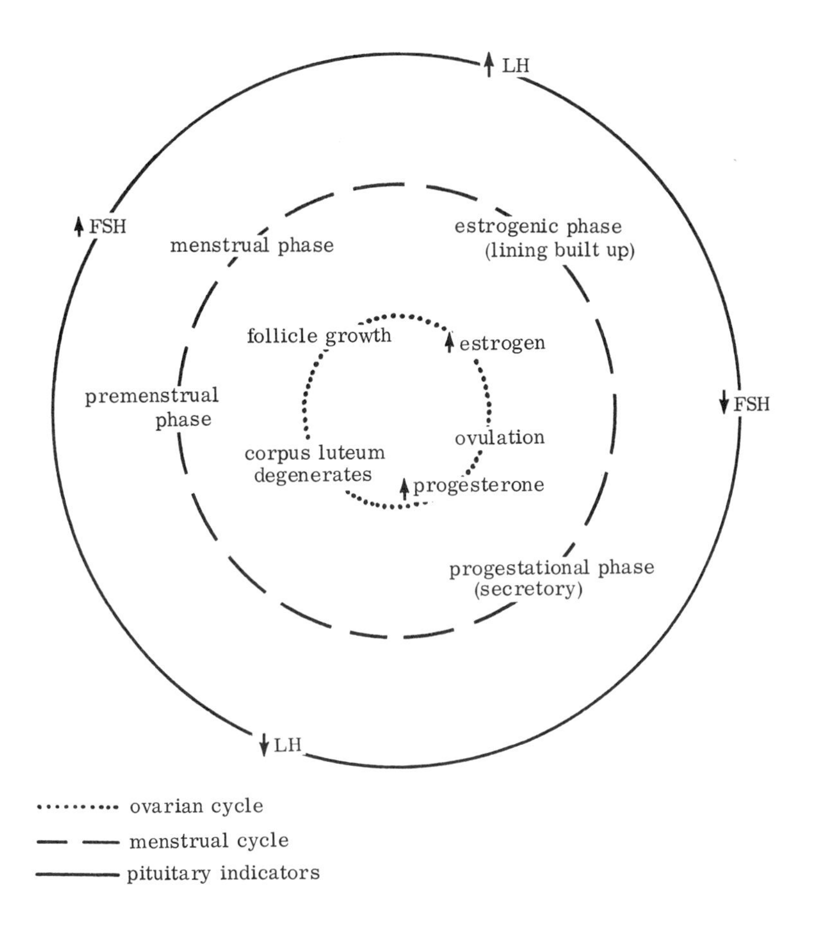

(a) What is the result of degeneration of a corpus luteum? ____________

____

(b) What hormone(s) is(are) secreted by the corpus luteum during pregnancy? ____________________

(c) Which phase of the menstrual cycle would you expect to find during pregnancy? ____________________

- - - - - - - - - - - - - - - - - -

(a) a corpus albicans (or decreased progesterone); (b) progesterone and some estrogen; (c) progestational (secretory)

BREASTS

15. The breasts, or mammary glands, are actually modified sweat glands. Their development at puberty is under the control of estrogen secretion. Estrogen and progesterone bring these structures to their fullest development late in pregnancy. The lactogenic hormone of the pituitary induces and maintains milk production or lactation. When milk ceases to be removed from the gland, its lobules stop functioning, and production ceases. Within each breast are 15 to 20 lobules, each with a duct leading to the nipple. The colored area surrounding the nipple is the areola. The areola becomes dark brown during pregnancy.

    The breast tissue itself is adjacent to the epimysium of the pectoralis major muscle of the chest wall. The breasts are physiologically connected with the rest of the female reproductive system. Breasts often vary in size or sensation with different phases of the menstrual cycle. Suckling a child causes the uterus to contract. And, normally, a woman does not resume her menstrual cycle while she is nursing a child.

    (a) What is the source and hormone that induce milk production?

    ____

    (b) Name the source and hormone that initiate breast development at puberty. ______________________________

    (c) Give two examples of the interrelationship of breasts with the reproductive system. ______________________________

- - - - - - - - - - - - - - - - - -

(a) pituitary, lactogenic hormone
(b) ovary, estrogen
(c) sensations, size variation, uterine contraction, no cycle while nursing (any two)

Menstruation

Menstruation is a normal, regularly occurring event in the lives of most women. Amenorrhea is the absence of menstrual flow. It may be temporary, reflecting some change in routine. It may reflect a glandular disorder or even pregnancy. Dysmenorrhea is painful menstruation and is abnormal. The cramps of menstruation may reflect poor posture, insufficient exercise, or a psychological problem. Many women experience mental depression, backache, and headache during the premenstrual phase. These may result from such factors as a tendency for body tissues to accumulate water during this hormonal stage.

Ectopic pregnancy

An ovum is normally fertilized in the uterine tube and implanted in the uterus. Occasionally implantation takes place in the tube, resulting in an ectopic pregnancy. The tubal epithelium is not designed to support implantation, nor can the muscular wall of the tube stretch to any great extent. Such a pregnancy is usually surgically removed before rupture occurs.

Birth control pills

In order for conception to occur, a woman must ovulate. Birth control pills (oral contraceptives) act by causing the body to suppress ovulation. Each pill contains enough estrogen and/or progesterone to prevent the secretion of FSH and LH to any great extent. Thus, there is no stimulus for a follicle to mature and develop into an ovum. Synthetic (artificially produced) hormones are used in birth control pills; since these can cause undesirable side effects, medical supervision is needed while a woman takes these pills over a period of time.

SELF-TEST

This Self-Test is designed to show how well you have mastered this chapter's objectives. Answer each question to the best of your ability. Correct answers and review instructions are given at the end of the test.

1. Which organ in the female reproductive system has a function similar to that of the vas deferens in the male? ______________________

2. Which layer of the uterine wall is the thickest? ______________

3. What event causes a corpus luteum to develop in the ovary? __________ ______________________

4. An area in the ovary consists of a sphere of stratified cuboidal cells surrounding a small fluid-filled cavity and a cell with a large nucleus. Name the area. ______________________

5. Name three components that help transport an ovum from the ovarian area to the uterus. ______________________

6. What are the direct effects of the two pituitary hormones on the ovarian cycle?

   FSH ______________________

   LH ______________________

7. During the estrogenic or proliferative phase of the menstrual cycle, what is taking place in the ovarian cycle? ______________________ ______________________

8. Very thick endometrium with much glandular activity is a sign of which phase of the menstrual cycle? ______________________

9. How does presence or absence of pregnancy affect the duration of the corpus luteum? ______________________

10. After ovulation, what are the normal times and sites of fertilization and implantation?

    Fertilization ______________________

    Implantation ______________________

Answers

Compare your answers to the Self-Test questions with those answers given below. If all of your answers are correct, you are ready to go on to the next chapter. If you missed any, review the frames indicated in parentheses following the answers. If you missed several questions, you should probably reread the entire chapter carefully.

1. uterine tube (Fallopian tube) (frame 6)
2. myometrium, smooth muscle (frame 5)
3. ovulation (frame 10)
4. secondary follicle or secondary oocyte (frame 8)
5. fimbriae, cilia, and peristalsis of smooth muscle (frame 6)
6. FSH—begins follicle maturation (increases estrogen); LH—converts collapsed follicle into corpus luteum gland (increases progesterone) (frame 12)
7. final maturation and ovulation (frame 14)
8. progestational or secretory (frame 8)
9. corpus luteum lasts one week if pregnancy does not occur, nine months if it does (frame 10)
10. fertilization within 24 hours of ovulation in uterine tube; implantation by a week later in uterus (frame 11)

Final Examination

Review and reinforce what you have learned about human anatomy in this book by taking this final examination. Answers are given following the test.

1. What two words describe the location of the shoulder in relation to the navel? ______________________________

2. What plane of section of the body would enable one to note the location of the stomach, liver, and brain? ______________________________

3. Name three major organs found within the thoracic cavity. ______________________________

4. Classify the tissue diagrammed at the right.

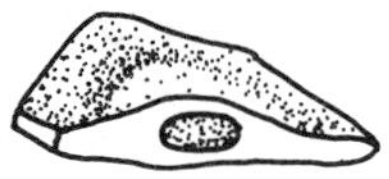

5. During mitosis the scattered chromatin forms 23 distinct pairs of chromosomes. During which phase does this occur? ______________________________

6. Name the structures that hold epithelial cells together. ______________________________

7. Name two cell types found in connective tissue and give the major function of each. ______________________________

8. Which functions of the skeletal system are provided by bones such as the femur? ______________________________

9. Identify the bones indicated in the diagram at the right.

   A ____________________

   B ____________________

   C ____________________

   D ____________________

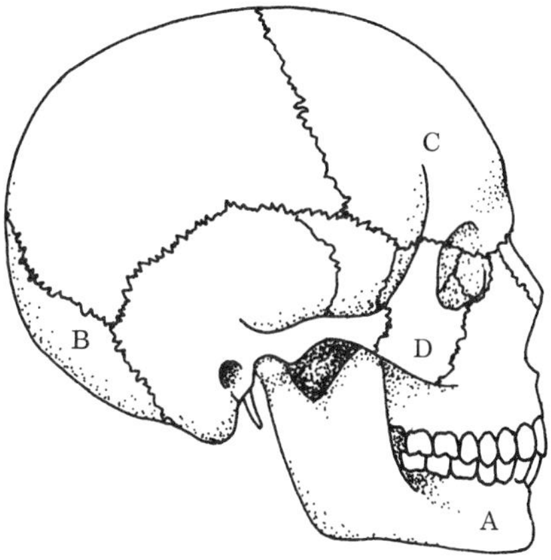

10. Name the bones that make up the following.

    (a) right pectoral girdle ____________________

    (b) thorax ____________________

    (c) toes ____________________

11. What muscle action is involved in bending the elbow? ____________________

    ____

12. What organelle in muscle cells is primarily involved in the contraction process? ____________________

13. The sartorius muscle extends from the os coxa to the tibia. Where is its insertion? ____________________

14. (a) Which letter indicates the area of the brain that receives sensory signals? __________

    (b) Which letter indicates the area that regulates and controls muscular coordination? __________

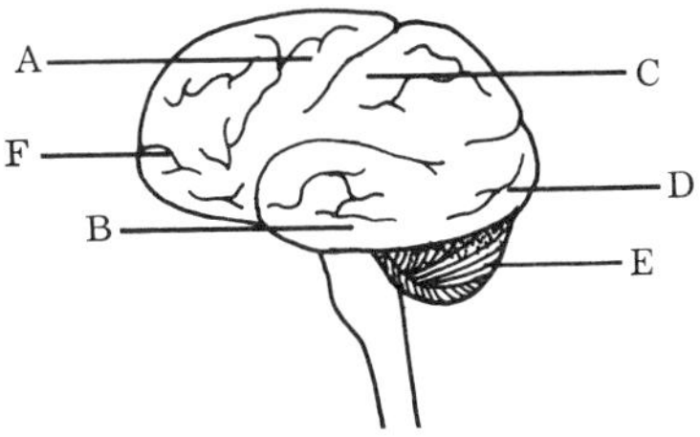

15. Name five elements present in each reflex arc. ____________________

    ____

    ____

16. A motor end plate involves components in which two systems? ________

____

17. Suppose an individual has no cones in the retina of the right eye. How would this affect his or her vision? ____________________

18. Receptors in the inner ear receive stimuli for which two types of sensation? ____________________

19. A free nerve ending in the skin produces what sensation when stimulated?

____

20. Certain pituitary hormones are produced in a portion of the central nervous system. How do these hormones reach the neurohypophysis where they are stored? ____________________

21. Explain the mechanism that controls thyroxin secretion. __________

____

____

22. With which other endocrine gland do the parathyroid glands control calcium metabolism? ____________________

23. How would you classify a formed element found in blood that has a segmented nucleus and many pink and blue stained granules? __________

____

24. Name the layer of a wall of a vein that is normally thickest. ________

____

25. Oxygenated blood from the lungs enters the heart through which vessel? ____________________ Into which chamber? __________

26. What is the function of the cluster of lymphoid structures found in each axillary region? ____________________

27. Through which lymphatic duct does lymph from the legs return to the systemic circulation? ____________________

28. What function do the thymus and the tonsils have in common? ________

____

29. Describe the epithelial lining of the trachea. ________________

____

30. The presence of what structures differentiates between terminal and respiratory bronchioles? ________________

31. Name the two nerves that supply the major muscles of respiration.

____

32. Name three specializations of the lining of the small intestine that increase absorption. ________________

____

33. In which digestive organ is each of the following located?
    (a) taenia coli ________________
    (b) pyloric portion ________________
    (c) jejunum ________________

34. List three functions of the liver. ________________

____

35. Which accessory digestive gland has both exocrine and endocrine functions? ________________

36. Which layers of the epidermis are modified when a fingernail is produced? ________________

37. Name the two layers of the dermis and give a distinguishing feature of each. ________________

38. List two structures used by the skin in accomplishing its function of temperature regulation. ________________

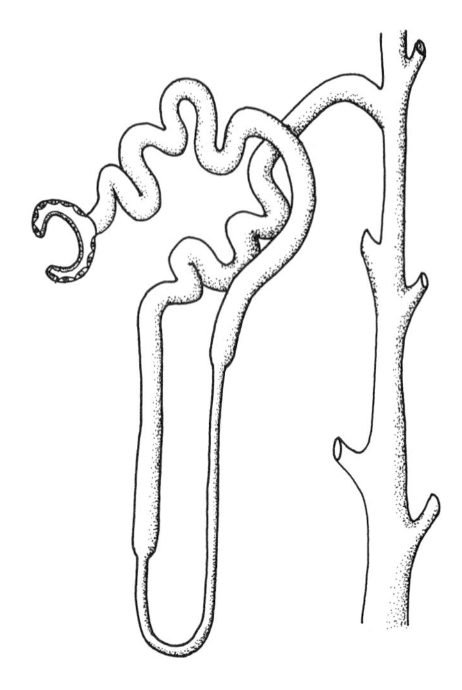

39. Label the following structures on the diagram above.

    A—glomerulus; B—proximal convoluted tubule; C—collecting tubule.

40. In what structure within the kidney does the filtration process occur?

    ____

41. What epithelial specialization lines kidney tubules? ______________

    ____

42. What four different systems are involved in excreting wastes from the body? ______________________________

43. In what structure are spermatozoa stored between their maturation and release from the body? ______________________________

44. What is the source of most of the fluid in semen? ______________

45. During development of spermatozoa, a cell division takes place in which the chromosomes are divided in half.

    (a) Name the type of cell division. ______________________________

    (b) Name the stage in development at which this occurs. ______________

    ____

46. (a) What substance is produced by interstitial cells in the testis?

____

(b) What controls the production of this substance? ______________

____

47. At the time of ovulation, what phase of the menstrual cycle would be in progress? ______________________________

48. What hormone is produced by the corpus luteum in a non-pregnant woman? ______________________________

49. Name the organs adjacent to the vaginal wall:

(a) anteriorly __________________

(b) posteriorly __________________

50. In relation to the ovarian cycle, what would be the ideal time for fertilization if pregnancy is desired? ____________________________

ANSWERS

Compare your answers to the final examination with the correct answers given below. If all of your answers are correct, congratulations! If you missed any questions, you may wish to review the chapter frames indicated in parentheses following the answers. (The number before the colon indicates the chapter; the number after the colon gives the appropriate frame number.)

1. superior, lateral (1:3-4)
2. coronal section (1:6)
3. heart, left lung, right lung (1:8)
4. squamous epithelium (2:8)
5. prophase (2:5)
6. desmosomes (2:7)
7. fibroblast—produces fibers; macrophage—digests particles; adipose—stores fat (any two) (2:12)
8. support, protection, blood formation, and mineral reserve (3:1)
9. A—mandible; B—occipital; C—frontal; D—zygomatic (3:7-8)
10. (a) clavicle, sternum (3:15)
    (b) vertebrae, ribs, sternum (3:13)
    (c) phalanges (3:24)

11. flexion (4:5)
12. myofibrils (4:3)
13. tibia (4:4, 4:9)
14. (a) A (5:12)
    (b) E (5:11)
15. receptor, afferent neuron, interneuron in spinal cord, efferent neuron, effector organ (5:8)
16. muscular and nervous systems (5:5)
17. poor color vision and poor vision in bright light (6:5)
18. hearing and balance (equilibium) (6:12)
19. pain (6:2)
20. they travel through axons (7:8)
21. negative feedback—TSH secreted by the adenohypophysis causes increased thyroxin secretion, which in turn decreases TSH secretion (7:3)
22. thyroid (7:12)
23. granular leukocyte or neutrophil (8:7)
24. tunica adventitia (8:17)
25. through the pulmonary vein; into the left atrium (8:22)
26. to filter lymph, to add immunity factors (9:7)
27. thoracic duct (9:4)
28. they both produce lymphocytes (9:12-13)
29. pseudostratified columnar with cilia and goblet cells (10:6)
30. alveoli (10:9)
31. intercostal nerve and phrenic nerve (10:16)
32. microvilli, villi, and plicae circulares (11:13)
33. (a) large intestine (11:17)
    (b) stomach (11:9)
    (c) small intestine (11:12)
34. production of bile, detoxification of blood, destruction of red blood cells, and phagocytosis (any three) (11:20)
35. pancreas (10:21)
36. keratinized and lucid layers (12:8)
37. papillary layer—thin connective tissue with papillae; reticular layer—thicker, coarse, or dense connective tissue (12:3)
38. sweat glands and capillaries in papillary layer (12:4)

39.

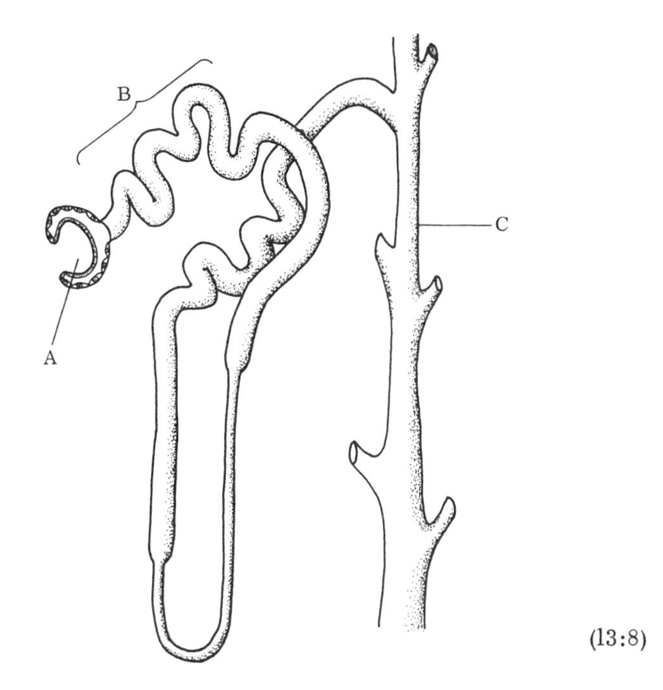

(13:8)

40. renal corpuscle or capsule (13:6)

41. microvilli for absorption (13:8)

42. respiratory, skin, digestive, urinary or excretory (13:1-2)

43. epididymis (14:6)

44. seminal vesicles (14:8)

45. (a) meiosis; (b) secondary spermatocyte (14:3)

46. (a) testosterone; (b) ICSH from adenohypophysis (14:12)

47. secretory phase (15:14)

48. progesterone (15:10)

49. (a) urethra; (b) rectum (15:3)

50. within 24 hours of ovulation (early progestational phase) (15:11)

Index and Pronunciation Guide